Comet Science

The Study of Remnants from the Birth of the Solar System

This book provides a comprehensive overview of our current knowledge of comets. It presents a fascinating survey of the study of comets throughout history, from antiquity to the present day, and includes the most recent discoveries on the exceptional comets Hale–Bopp and Hyakutake. The authors discuss the role of comets in the formation of our Solar System and describe the links between comets, asteroids and the recently discovered Kuiper-belt objects. The book also includes new insights into the composition and nature of cometary nuclei, with results from the most up-to-date observation techniques.

Written in a clear and lively style, and beautifully illustrated, this book will appeal to anyone interested in comets and astronomy, professionals and amateurs alike. It will be of particular interest to students and researchers in astronomy, astrophysics and planetary science, as well as general readers with a good background in physics.

Jacques Crovisier is an astronomer at the Paris-Meudon Observatory, specialising in the radio and infrared observation of comets and the modelling of molecular processes in cometary atmospheres. He is a co-investigator on the future Rosetta mission to comet Wirtanen.

Thérèse Encrenaz is the Director of the Department of Space Research (DESPA) at the Paris-Meudon Observatory. She is a specialist in the remote sensing of planetary atmospheres. She has been involved in several space missions, including the International Halley Watch, and is co-investigator for many future missions, including Rosetta.

Jacques Crovisier and Thérèse Encrenaz
Paris-Meudon Observatory

Translated by Stephen Lyle

Comet Science

The Study of Remnants from the Birth of the Solar System

Foreword by Roger Maurice Bonnet

PUBLISHED BY THE PRESS SYNDICATE OF THE UNIVERSITY OF CAMBRIDGE
The Pitt Building, Trumpington Street, Cambridge, United Kingdom

CAMBRIDGE UNIVERSITY PRESS
The Edinburgh Building, Cambridge CB2 2RU, UK www.cup.cam.ac.uk
40 West 20th Street, New York, NY 10011-4211, USA www.cup.org
10 Stamford Road, Oakleigh, Melbourne 3166, Australia
Ruiz de Alarcón 13, 28014 Madrid, Spain

English edition first published 2000

Printed in the United Kingdom at the University Press, Cambridge

Typeface Caslon MT 10/14pt, by Keyword Typesetting Services Ltd

A catalogue record for this book is available from the British Library

Library of Congress Cataloguing in Publication data

Crovisier, Jacques, 1948–
[Comètes. English]
Comet science : the study of remnants from the birth of the solar system / Jacques Crovisier and Thérèse Encrenaz ; translated by Stephen Lyle.
p. cm.
Includes index.
ISBN 0 521 64179 9 (hc.). – ISBN 0 521 64591 3 (pbk.)
1. Comets. I. Encrenaz, Thérèse, 1946– . II. Title
QB721.C7613 2000
523.6–dc21 99-19612 CIP

ISBN 0 521 64179 9 hardback
ISBN 0 521 64591 3 paperback

Contents

Preface to the English edition

At the end of July 1995, at the very moment the first (French) edition of this book was in the press, comet Hale–Bopp was discovered. In Spring 1996, comet Hyakutake made a close approach to Earth. 1996 and 1997 were two hectic years for cometary scientists with the observations of these two exceptional comets. A wealth of new results emerged. For example, the number of molecules shown to be outgassed from cometary ices was multiplied by three.

Observations from space cast new light on cometary science. The Infrared Space Observatory unravelled the infrared spectrum of comets. X-ray satellites unexpectedly found out that several comets were significant sources of X-rays. The coronagraph aboard the SOHO solar observatory discovered no fewer than sixty comets when they grazed the Sun.

As a matter of fact, most of our previous ideas on the nature of comets successfully faced the challenge of these new observations. The basic concepts were confirmed. But new results led to an in-depth assessment of the classical ideas, revealing solutions to some old problems, and the emergence of several new enigmas.

The present edition has been updated and expanded to try to give a full account of these new developments.

Jacques Crovisier
Thérèse Encrenaz
Meudon, 1999

Foreword to the French edition

Once Halley's comet had turned away from us, leaving various probes scattered across space after their spectacular encounter with the comet of all comets in March 1986, the frenetic excitement of the scientific community appeared to give way to its usual Aristotelian order, as if inspired by the celestial object which had so suddenly provoked it. However, the book by Jacques Crovisier and Thérèse Encrenaz has come just at the right moment to dispel any illusion of calm. Cometary research has not fallen back into the slumbers of an unfashionable discipline, as it might have been described before 1986, too specialised to mobilise any wide-ranging research effort. Today cometary science has been raised to a status befitting its true preoccupation: the formation of the Solar System and the origins of life. Halley has caused more than one revolution. In 1705, it was the astronomer Halley who identified the comet and predicted its periodic return, thereby producing brilliant confirmation of Newton's theory of gravitation. In 1986, it was comet Halley, on its return to perihelion at the height of the space age, exposing itself to the extraordinary advances in astronomical detection and imaging.

This great step has now been made. The eleven chapters of the present work give a precise and I think complete description of comets as we now know them; three long chapters present a clear and full explanation of the nature of comets. Dishevelled and bloodless, pallid carriers of ill omen, comets were for years the very symbol of an irrationalism which has not entirely disappeared. Today they are making a noble return to fame, entirely justified, as this book proves without difficulty.

Where do they come from? From the Oort cloud, tens of thousands of astronomical units away? From the Kuiper belt, beyond the orbit of Neptune? It is particularly moving to read the authors' comments on these ideas, just as the Hubble Space Telescope is beginning to confirm the existence of such a belt. Comets are rather special members of the great family of small bodies, for there is little doubt that they are its most primitive and least evolved members; that they have directly witnessed the formation of the Solar System. Progress in observational astronomy and *in situ* analysis means that we can today compare cometary grains with interstellar dust, thereby tracing back to the very genesis of our Solar System. It turns out that comets resemble dirty snowballs, indeed, very dirty concoctions of ice and dust. Heated by the Sun, wreaths of gas and dust inseminate the interplanetary medium and any object they may encounter, including Earth, with astonishingly complex, almost prebiotic organic compounds.

They are a kind of astronomical spermatozoid! This distinguishes them from asteroids and meteorites, which would appear to be rather sterile, mineral nuclei, devoid of any icy envelope or other compounds. It is this which, at the peak of their activity, can make comets such spectacular objects.

We shall see that comets could have devastating effects on our planet if their trajectories should cross Earth's orbit. A whole chapter is devoted to the case of comet Shoemaker–Levy 9 which crossed Jupiter's orbit. It inflicted several tremendous blows to the giant planet, of energies hundreds of times greater than the whole nuclear arsenal on Earth, and left scars which lasted for several weeks.

There are also four chapters devoted to techniques of observation and analysis, with predictions as to how they will develop. Ground-based and orbital astronomy, ultraviolet and visible spectroscopy, and in particular, infrared and radio spectroscopy are all covered. Also exposed is the crucial role played by technical advances which, surprisingly enough, remain

European Southern Observatory (ESO) at La Silla, Chile. The site is located in the Andes at altitude 2400 m. ESO document.

solely in the hands of European astronomers, after the success of the Giotto probe and with preparations already underway for the Rosetta mission, due to be launched towards Wirtanen in 2003. Europeans have also been at the forefront of technological advance with two infrared missions, ISO and FIRST, both of which included comets amongst their primary targets.

Who should read this book? Certainly, any student or specialised research scientist. Its clear style is a perfect vehicle for the high quality of its explanations. Mechanisms of cometary interaction, activation and formation are laid bare in a remarkable and simple way. The language used is not over-specialised, nor weighed down by complicated formulas. Readers will appreciate the happy balance between chapters devoted to observation and those devoted to techniques, as well as the high quality and conciseness of explanatory boxes separated from the main text. In brief, everything you have always wanted to know about these objects and never dared to ask, or have not yet understood, you are likely to find here. You will also find all those questions awaiting elucidation. So I will not tempt your curiosity further.

Roger Maurice Bonnet
Director of Scientific Programmes, ESA
1995

1 The history and science of comets

1.1 The first observations

Comets have been known to mankind since ancient times. Even in the time of the Chaldeans four thousand years ago, astronomers were struck by the unpredictable, changeable and sometimes spectacular aspect of these objects. Like the planets, which were referred to as 'wandering stars' by the Greeks, it was only natural that comets would attract attention by their motion relative to other stars. In addition to this, their sudden changes in brightness, the appearance of trails of light associated with them, and their rapid and apparently inexplicable movement across the sky could only cause consternation and amazement in the minds of their observers. They found themselves quite unable to understand their true nature or behaviour. The Egyptians gave them the name 'comet', or 'hairy star'. Together with the 'head', or central part of the body, and the 'tail', a rectilinear area of light, this 'hair' was indeed their most noticeable feature.

It is human nature to fear what is not understood; but, rather than admit their ignorance, astrologers preferred to attribute a divine origin to those events they were unable to explain. It was thus that a myth developed, according to which comets were messengers from beyond, announcing ill omens. This superstition persisted for centuries until, in the eighteenth century, the astronomer Edmond Halley demonstrated the true nature of cometary orbits. However, history has shown that the consequences of this myth remained present well after that date, and have not yet been entirely exorcised.

Figure 1.1. opposite
Comet West 1976 VI on 12 March 1976, with its tail 100 million kilometres long. Photograph by S. Koutchmy, courtesy of CNRS-IAP.

1.1.1 Aristotle's mistake and its consequences

It is known that the Greeks, adhering to the theories of Aristotle from the fourth century BC, and then of Ptolemy two centuries later, considered the Earth to be at the centre of the Universe. A bold hypothesis put forward by Aristarchos of Samos, according to which the planets were all moving around the Sun, was largely ignored in his day (around 250 BC). It lay forgotten for over a thousand years, until the work of Copernicus in the sixteenth century finally reasserted the heliocentric system.

In order to account for planetary motions, Ptolemy had assumed that each one of them followed a small circular path, or epicycle, whose centre itself moved along a circular path around the Earth. With the complications that this

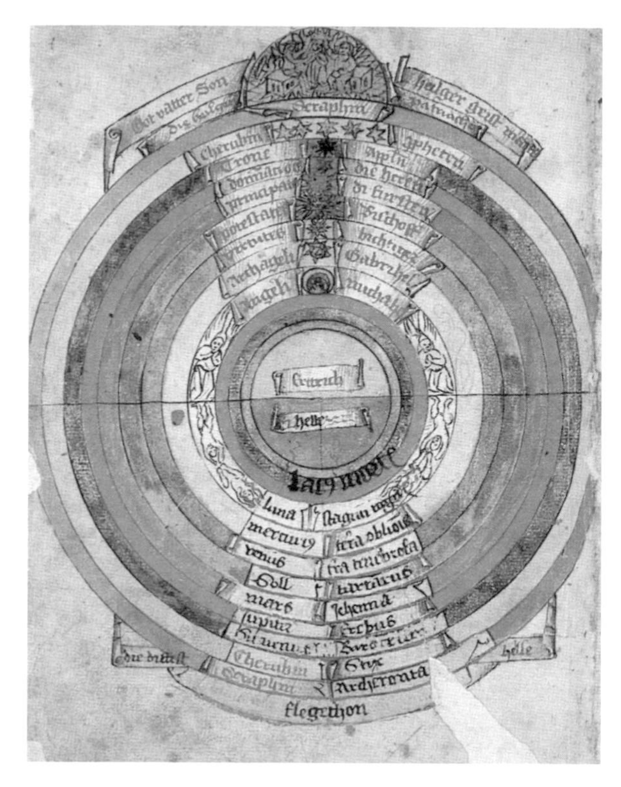

Figure 1.2.
The pre-Copernican geocentric Universe. The Earth, at the centre, is enclosed by a series of three layers representing water, air and fire, followed by more distant concentric circles representing in turn the Moon, the seven planets known at the time, and the Sun. German manuscript, *circa* 1450.

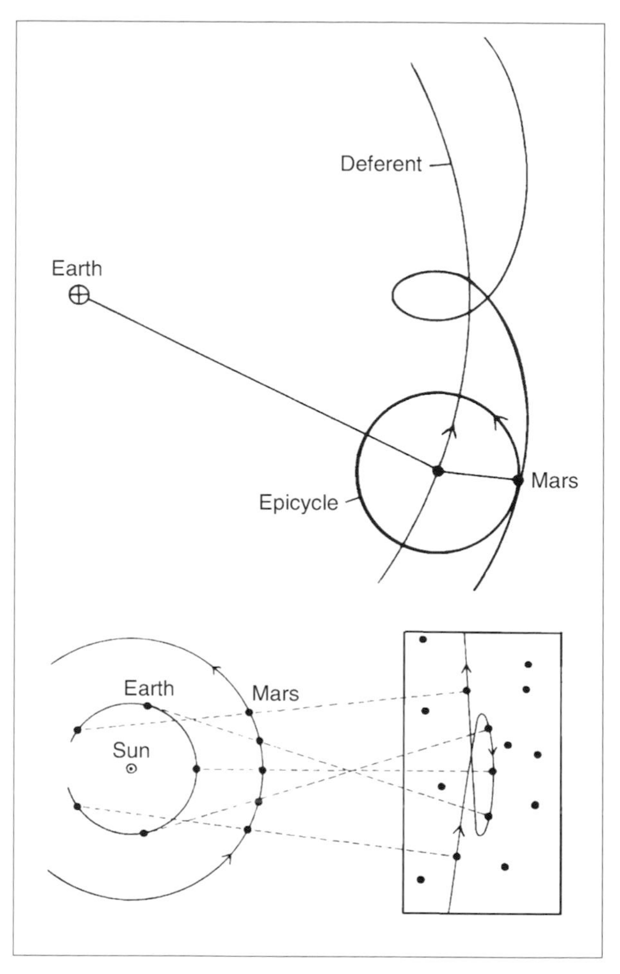

Figure 1.3.
The Solar System, according to Ptolemy and Copernicus. In Ptolemy's geocentric system, the planet is located on a small circle (an epicycle) which is itself moving along a larger circle (the deferent) centred on the Earth. In the heliocentric system of Copernicus, the planets follow circular orbits centred on the Sun. From Sagan, 1981 [74].

method introduced, astronomers of the time were able to explain planetary orbits. But what of the comets? Aristotle, finding himself unable to understand them as interplanetary bodies, assigned them an atmospheric origin. This belief, which was to persist for over twenty centuries, did more than anything else to convince people that there was a relation of cause and effect between cometary apparitions and certain terrestrial phenomena, whether they were of natural origin, such as floods, earthquakes and plague, or human, such as war.

This idea predominated right through from ancient times to the Renaissance. In this context, the unfortunately isolated voice of Seneca (4 BC to AD 65) in the first century AD seems all the more extraordinary to us today. He was indeed the first to suggest that comets might be periodic. Beyond this, he opened the way to a truly scientific approach, based upon observation and stripped of prejudice, as is illustrated by the following quotation from his *Natural Questions*:

Why be surprised that comets, which so rarely reveal themselves to the world, should not yet be subject to our

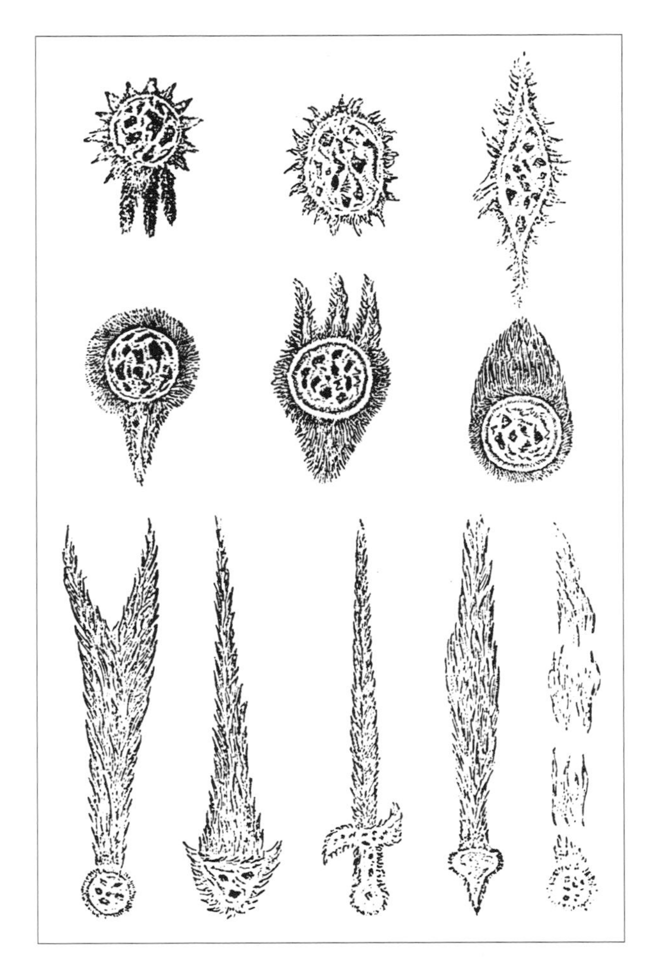

Figure 1.4.
Various cometary shapes recorded by Pliny the Elder. Engraving taken from the *Cometography* by Hevelius (1668).

rigid laws, and that we know neither the origins nor the destinations of these bodies, whose passages are separated by such immense intervals? The time will come when a careful study, pursued over several centuries, will throw light on these natural phenomena. A man will one day be born who will show us over which part of the sky these comets wander, why their motion is so great in contrast to the planets, and what is their size and nature.

This man was indeed born, eighteen centuries later. His name was Edmond Halley, and his discoveries were to confirm the brilliant intuition of the Roman savant.

Seneca's suggestions found no echo in his time, nor for many generations afterwards. His contemporary Pliny the Elder (AD 23 to 79) devoted a purely descriptive work to comets, in which he restated Aristotle's 'atmospheric' theory. This laid the way open to superstition.

Whereas the Greeks interpreted comets as apparitions to ordinary mortals of the gods of Olympus, the Romans saw in them a presage of ill omen. It was thus that Nero had several dignitaries assassinated, in order to escape from the curse associated with the passage of a comet, in fact the future Halley's comet. Later, apparitions of comets were said to have announced the death of Attila in 453, of the emperor Valentinian in 455, of Merovee in 457, of Chilperic I King of the Franks in 584, and of Muhammad in 632, among many others.

The comet of AD 1000, whose apparition coincided with an earthquake, caused even greater consternation; and in 1066 a further apparition of what was to become Halley's comet coincided with the death of Harold II, defeated by William the Conqueror at the battle of Hastings. The scene is immortalised on a now famous section of the tapestry at Bayeux. During the following centuries, comets continued to inspire terror, notably in 1402 and 1577.

In 1301, on the other hand, the Florentine painter Giotto was so impressed by a cometary apparition, which was once again none other than Halley's comet, that he portrayed it in the place of the Star of Bethlehem, in his painting *The Adoration of the Magi*. This is one of the rare cases in which a comet has been associated with a joyful event. Some have suggested that it was Halley's comet itself, during a passage in AD 12, which could have produced the exceptional brightness of the Star of Bethlehem, although the most popular hypothesis at present is a conjunction of Jupiter and Saturn in 7 BC.

Figure 1.5.
Halley's comet on the Bayeux Tapestry. This fragment of the tapestry shows the consternation of Harold II and his court when they saw the comet in 1066. Some time later, Harold was beaten by William the Conqueror at the battle of Hastings. By special permission of the city of Bayeux.

1.1.2 The advent of cometary physics

Throughout the Middle Ages cometary physics made little progress. The first astronomical measurements were made in the fifteenth century, in Germany and Italy. Regiomontanus (1436–1476) undertook systematic measurements of cometary motions. Frascator (1483–1553) suggested that cometary tails were always directed away from the Sun, and this was confirmed by the observations of Pierre Apian (1495–1552). A century later, Tycho Brahe (1546–1601) was the first to determine the distance of a bright comet, which appeared in 1577, by simultaneously observing it from two observatories separated by a distance of 600 km. He concluded that the comet had to be located much further away than the Moon. This constituted a decisive blow to the 'atmospheric' theory of comets.

The sixteenth century was also to see the advent of the Copernican system, with the publication by Copernicus (1473–1543) of his famous work *De Revolutionibus Orbium Celestium* in 1543. It was to be several decades before this revolutionary theory was accepted, notably due to obscurantist opposition from the Church. A further great step for astronomy was taken in 1609, when Kepler (1571–1630) formulated his laws governing planetary orbits around the Sun, in his work *Astronomia Nova*. This was to be the final blow to the Ptolemaic system.

The scene was then set for Edmond Halley's great discovery (1656–1742): comets follow highly elliptical orbits around the Sun, and thereby make periodic close approaches to both the Sun and Earth. The work of this English astronomer was based partly on many observations of past comets, and partly on the universal theory of gravitational attraction recently propounded by Isaac Newton (1642–1727), and published in his *Principia* in 1687. In 1705, Halley published the theory which was to bring

Figure 1.6.
The Adoration of the Magi, painted by Giotto di Bondone on the walls of the Scrovegni chapel in Padua, around 1301. The comet at the top of the scene may well be Halley's comet, the fresco being contemporary with one of its passages. This painting inspired the name for the European space probe Giotto.

him fame, under the title *Astronomiae Cometicae Synopsis*. He asserted that the comet observed in 1456 was the same as that observed in 1531, 1607 and 1682. As the period was 76 years, he announced its return for 1758, and indeed on 25 December of that year the comet was once again observed, to the glory, albeit posthumous, of its discoverer. At the same time, Newton's theory had been brilliantly confirmed and the foundations of cometary physics firmly established.

Once the period of Halley's comet had been determined, astronomers and historians began to look for its previous apparitions in ancient records. Their efforts were richly rewarded.

Figure 1.7.
Aztec painting showing the emperor Montezuma observing a comet. The Aztecs associated sinister prophesy with such events. Courtesy of D. Duran, *Historia de las Indias de Nueva España.*

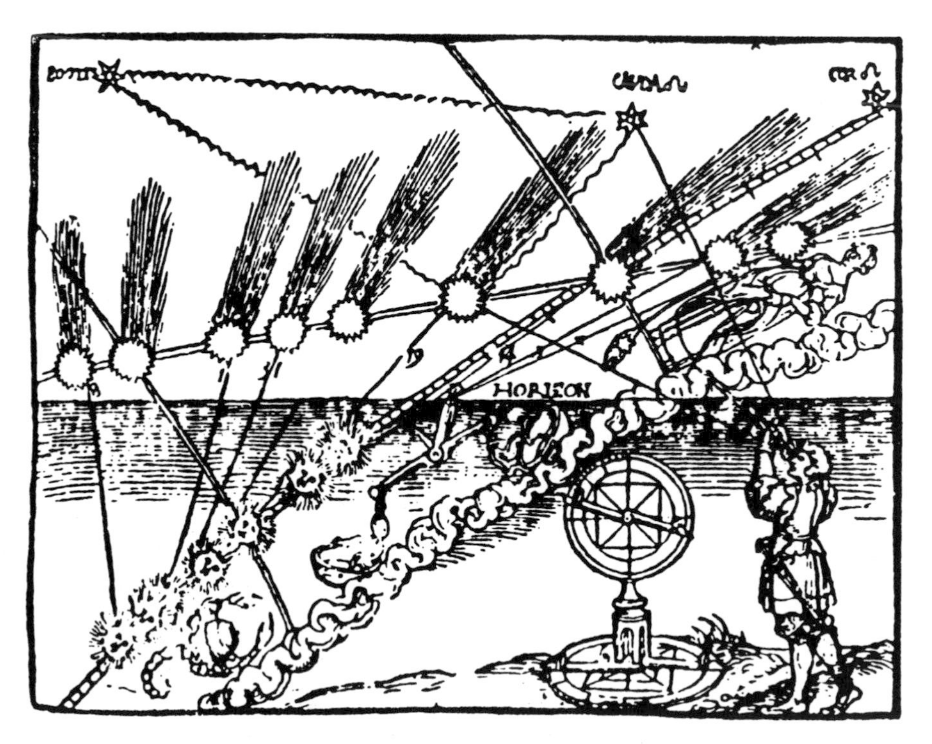

Figure 1.8.
The observations of Pierre Apian (1495–1552) showed that the cometary tail is always directed away from the Sun. Courtesy of the Observatoire de Paris.

Figure 1.9.
The English astronomer Edmond Halley (1656–1742), who observed in 1681 and 1682 the comet to which he was to give his name. He calculated its orbit and confirmed Newton's theories by correctly predicting its return in 1758. Courtesy of the Observatoire de Paris.

Since 240 BC, when the comet had been observed by Chinese astronomers, most of its apparitions had left some trace in the records of the time. Some of them had been quite spectacular. Apart from the examples mentioned above, the apparition in AD 451 should be noted, coinciding with the victory of the Romans and the Visigoths over Attila; as well as that of 1456, which the Church associated with the fall of Constantinople, taken by the Turks three years previously.

In 1759, after this brilliant confirmation of Halley's theory, it might be thought that the now identified comets would cease to terrify mankind. But this was not the case, for nothing at all was known of their size, nor of their true nature, and in view of their periodicity it remained a possiblity that one might actually collide with Earth. The French writer Molière mused in humouristic vein over this possibility in his *The Learned Ladies* ('We have had a narrow escape tonight, Madam. A world close to us has passed right by... '). But many were genuinely concerned, and controversy raged across the literary and scientific world. The comet of 1773 inspired widespread panic, despite scientists' reassurances of the low probability for such an event. Scenes of panic were repeated in 1798, 1816, 1832, 1857, 1872, and 1899. In 1910, Halley's comet once again moved the populace to the heights of emotion, this time in fear of highly toxic cyanogen gas from the cometary tail as it swept across Earth's atmosphere.

The last apparition of Halley's comet in 1986 has shown that, although comets no longer inspire the same terror, they nevertheless continue to fascinate the general public, in a way which goes beyond mere scientific curiosity. It could be said that the old superstitions have not been altogether eradicated.

1.2 In search of our origins

Paradoxically, it was at the very moment when comets were divested of their mythological aspect that their interest for the scientific world began to grow. Indeed, having been totally freed by Halley's discovery of the atmospheric character previously attributed to them, comets were recognised as bodies belonging to the Solar System, in just the same way as the planets and their satellites. For this reason, they deserved to be studied as such.

Many comets were observed in the nineteenth century. Among them, a comet of particularly short period, only 3.3 years, was closely

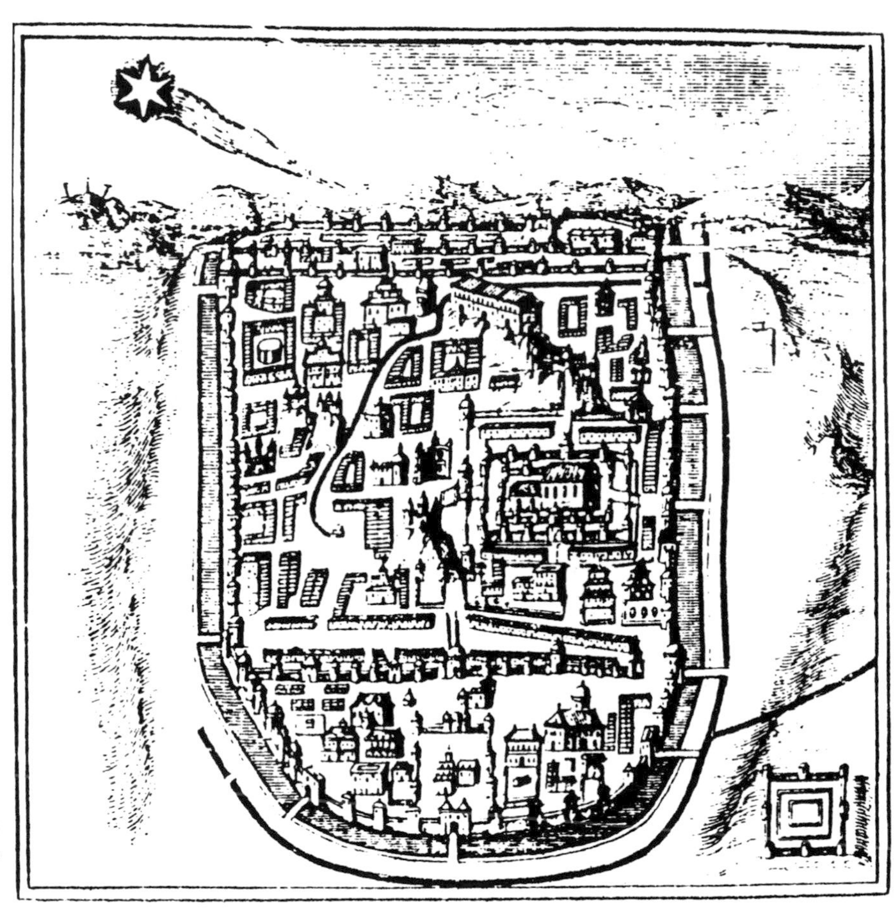

Figure 1.10.
The comet of AD 63 over Jerusalem. This was not Halley's comet, which appeared in AD 66. Courtesy of the International Halley Watch (IHW).

examined. This was comet Encke, observed for the first time in 1786, and named after the German astronomer Johann Franz Encke (1791–1865). Encke was the first to predict its return in 1822. The very short period of this comet made multiple observation possible, and Encke noticed that the period was not exactly that predicted by Newton's theory of gravity. In fact, it was about two and a half hours shorter. This slight difference could only be explained by the existence of another force, which was referred to as 'non-gravitational', its origin being unknown.

After Halley, the second great step in the history of cometary physics was made by the American astronomer Fred Whipple (born in 1906). In 1950, based on an analysis of the non-gravitational forces, Fred Whipple put forward a model to explain the nature of comets, that became famous under the name of the 'dirty snowball' hypothesis.

According to Whipple, a comet was not a diffuse aggregate of particles, as others had suggested before him, but rather a solid nucleus only a few kilometres across, composed mainly of water ice mixed with solid particles; whence the appellation 'dirty snowball'. When a comet is far from the Sun, its nucleus is too small to be observed by telescope. As it approaches the Sun, however, the side exposed to light is heated and the ice evaporates, carrying with it

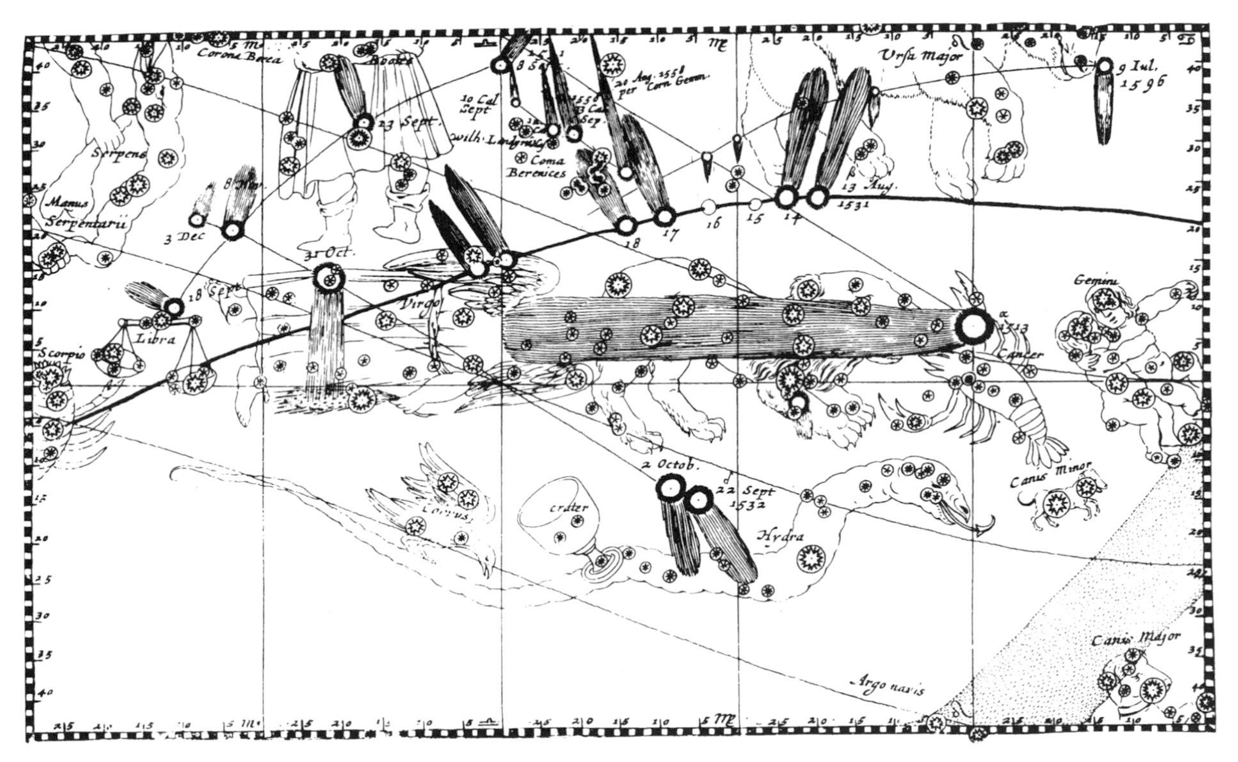

Figure 1.11.
Illustration of Halley's comet during its passage in 1531. Engraving taken from the work of Stanislas Lubienietz, *Theatrum Cometicum*, Volume 2, Amsterdam, 1667. Courtesy of the Observatoire de Paris.

any dust held within. A halo of dust is thereby formed around it, which reflects the light of the Sun and renders it visible from Earth. The sometimes spectacular variations in the appearance of comets as they follow their trajectories was thus explained. Those non-gravitational forces observed in certain comets are due to braking (or acceleration) as matter is ejected in the direction of the Sun.

It was not until 1986 that space exploration of Halley's comet could bring confirmation to Whipple's theory, although by then it had become so widely accepted that the result came as no surprise.

During the second half of the twentieth century, our knowledge of comets has gradually progressed and the great interest of these objects for the study of the Solar System as a whole has been revealed. For here are bodies which remain for most of their time at immense distances from the Sun. Some of them, such as Halley's comet, pass beyond the orbit of Neptune. Consequently, they are unlikely to have suffered any significant modification since the moment of their formation. Indeed the very weak gravity at such distances, particularly for objects of such small nucleus, would spare them any metamorphosis. Likewise, the low temperatures to which their nuclei are subjected mean that even the most volatile of their constituent molecules can remain in a solid state. The only change expected would be the loss by evaporation of an outer layer of ice and dust, each time they pass close to the Sun.

Figure 1.12.
Comet Bennett 1970 II over the Alps. IHW, photo by C. Nicollier.

Given that comets have evolved so little since their origin, they can be considered as an invaluable witness to the history of the Solar System. Moreover, such witnesses are rare. The terrestrial planets have evolved enormously, as is demonstrated by their great diversity. The Jovian planets are far away, and only their outer layers can be studied. As regards meteorites, these come from asteroids which, more often than not, have undergone numerous transformations. Hence the unique opportunity provided by these *primitive* objects, particularly since in certain circumstances they come close enough to Earth to be studied in great detail.

1.3 A unique laboratory

Apart from their interest as 'fossils', revealing the primitive composition of the Solar System, comets offer observers all the advantages of physicochemical conditions which differ radically from those prevailing in the terrestrial environment. For example, an extremely low temperature, of only a few dozen kelvins, within the nucleus; an extremely low pressure; an almost zero gravitational field; a transient cometary

Figure 1.13. opposite
Comet West 1976 VI in February 1976, photographed from Gromau in Germany. Photo by M. Grossman.

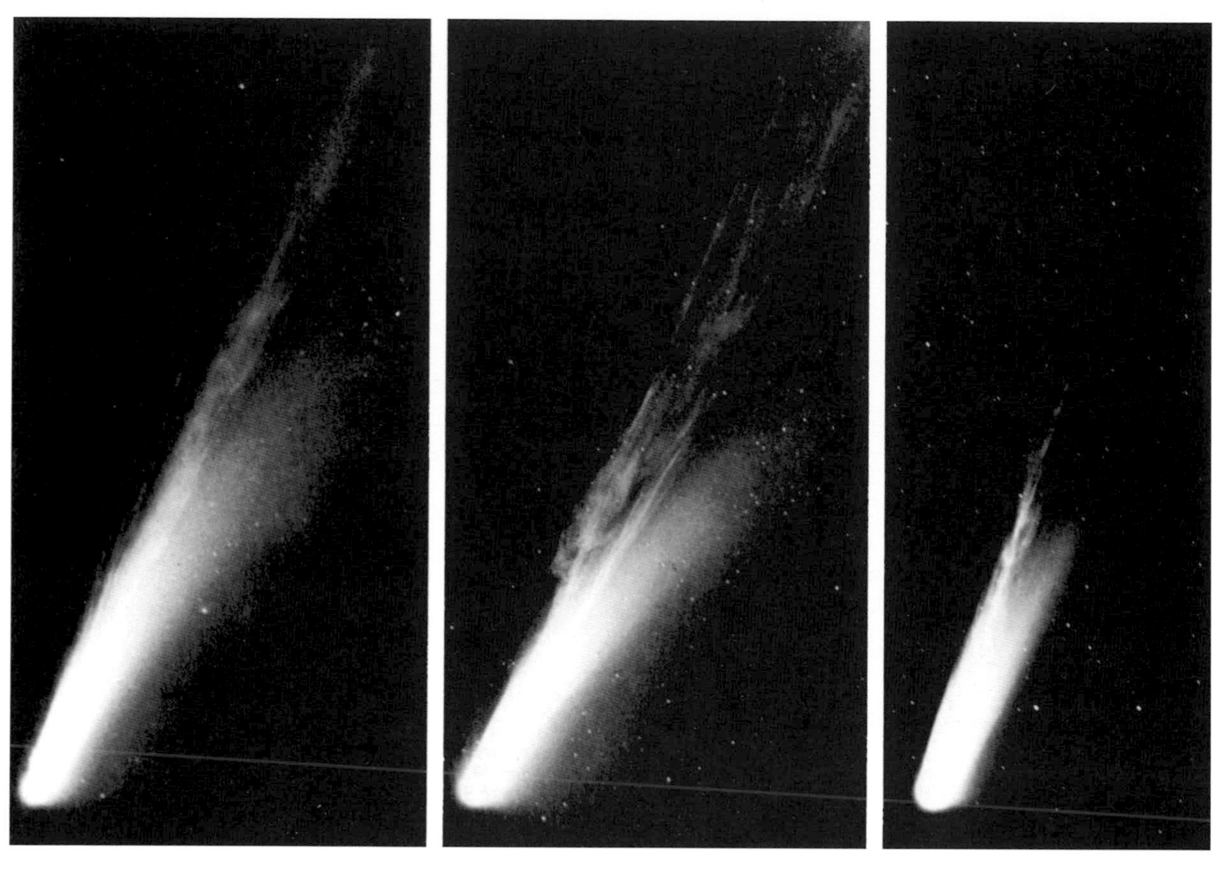

Figure 1.14.
Comet Mrkos 1957 V photographed on 17, 22 and 27 August 1957. Note the evolution of the ion tail, exhibiting filaments and turbulence, whilst the dust tail remains homogeneous. Photo Hale Observatories.

Figure 1.15.
Photograph of Halley's comet taken in May 1910. The particularly favourable geometric configuration of this encounter resulted in records of exceptional quality. In the *bottom left*, Venus gives an idea of the brightness of the comet. Photo by the Lowell Observatory, Flagstaff, Arizona, courtesy of the Observatoire de Paris-Meudon.

atmosphere, in a state of permanent escape, and subject only to solar radiation in the visible and ultraviolet, which dissociates and ionises its constituent molecules. Such are the typical conditions in the cometary environment.

These conditions are not easy to reproduce on our planet. Cometary physics can therefore teach us a great deal about the constitution of ices at low temperature, chemistry in a low density and cold gaseous phase, the physics of changes of state at low temperatures, heterogeneous chemistry and interactions between molecules in the gaseous state and dust grains, photochemistry, and molecular spectroscopy.

The comets have no doubt lost a certain part of their mystery, but at the same time they have gained considerably in scientific interest. Ultimate witnesses of the origin of the Solar System, and evolving in extreme physicochemical conditions, comets remain as fascinating as ever.

2 What is a comet?

For more than two centuries, it has been known that comets, like planets, are bodies subject to the solar gravitational field. They follow highly eccentric orbits which take them, in certain cases, to very great distances from the Sun, further even than the most distant of the Jovian planets. Moreover, the theory proposed by Fred Whipple, and confirmed recently, informs us of the essential features of these objects. They are small celestial bodies, several kilometres in diameter, constituted mainly of water ice and rock.

2.1 The nucleus, coma and tails

Far from the Sun, comets consist of a nucleus alone and, because this nucleus is neither very large nor very bright, they remain difficult to observe. But as the comet approaches the Sun, the surface temperature of its nucleus rises, causing sublimation of ices and ejection of gas and dust. The dust scatters light from the Sun, and this scattered radiation observed from Earth is known as the *coma*, from the Latin word for hair. The coma spreads out as the comet approaches the Sun. If it is sufficiently active, in the sense that gas and dust ejection is on a large enough scale, then two tails may form, one wide and curved, the other narrow and straight. The former is due to dust scattering solar light, and the latter is caused by ionised gases fluorescing under excitation from solar radiation.

Cometary activity depends on two factors: its intrinsic composition, in particular its content of volatile substances; and its distance from the Sun, activity increasing for orbits passing close to the Sun. The problem of observing phenomena associated with cometary activity depends on a third factor, namely the distance of the comet from Earth at the time of observation.

As mentioned earlier, water is the main constituent among the volatile substances. Knowing the energy required to cause sublimation of water, and the radiation emitted by the Sun, it becomes possible to calculate the heliocentric distance at which this sublimation is likely to take place. Calculations show that outgassing can be expected at a temperature of around 200 K, which corresponds to a heliocentric distance of some 2.5 AU. However, many comets prove to be active at much greater heliocentric distances, which implies the presence in the nucleus of other volatile compounds. This is exemplified by carbon monoxide (CO) and carbon dioxide (CO_2), both detected on comets Halley and Hale–Bopp. Other molecules, such as hydrogen cyanide (HCN) and methanol (CH_3OH), have also been detected, and will be described later.

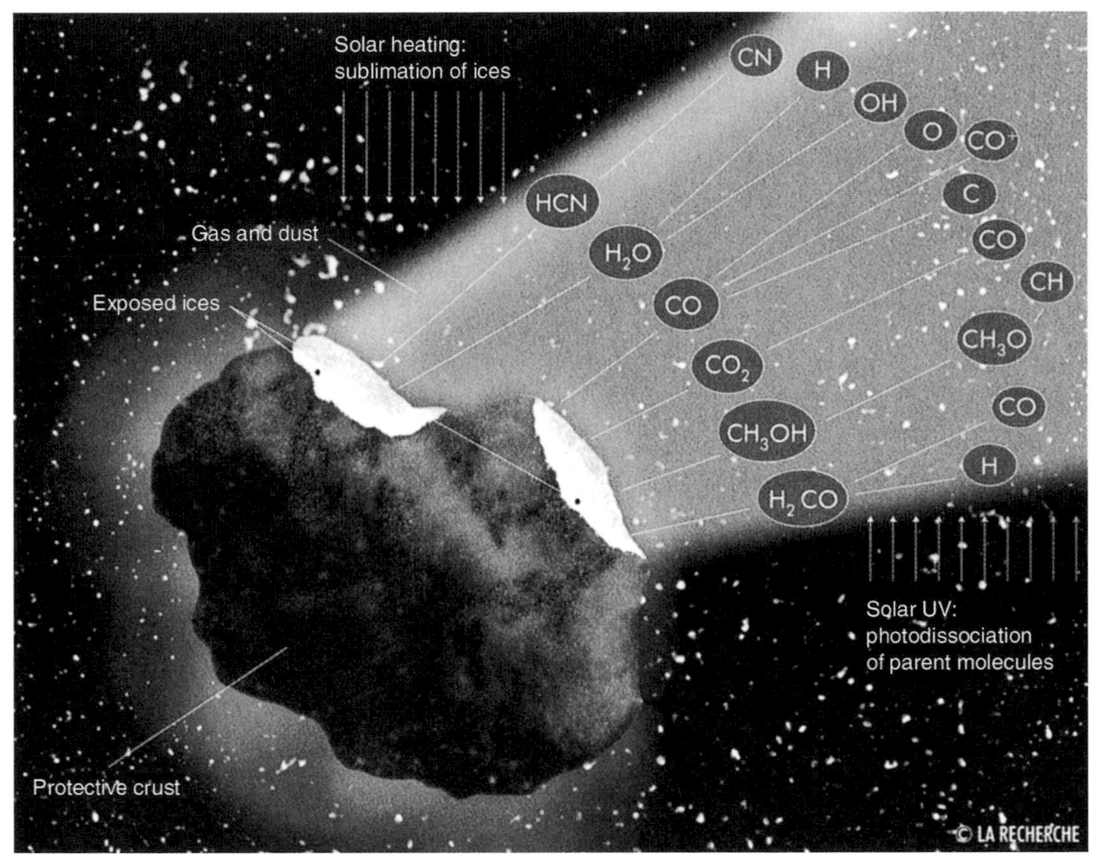

Figure 2.1.
The physical processes of cometary activity. Volatile elements in the nucleus sublime under solar heating. Such molecules are *parent molecules*, which bring with them dust contained in the nucleus. These molecules are progressively dissociated under the influence of ultraviolet radiation, forming *secondary products* or *daughter molecules*, which are, in fact, radicals, ions and atoms. Bockelée-Morvan and Crovisier, 1994 [30]. Copyright 1994, La Recherche, Paris, France.

Volatile constituents sublimed by solar radiation carry with them dust from the nucleus, thus forming a coma around the nucleus. However, parent molecules released in this way have only a relatively short lifetime. Under the action of ultraviolet solar radiation, they dissociate into radicals, ions and atoms. The dust trails behind in what is called a *dust tail*, whilst the ions line up in the antisolar direction, away from the Sun, in the *ion tail*, sometimes extending out over several million kilometres.

Hence, comets gain their characteristic appearance mainly by cometary particles scattering the visible solar flux. The nucleus, on the other hand, is largely obscured by the envelope of gas and dust it ejects. In order to understand the chemical composition of comets, that is, the nature of ices and rocks which comprise the nucleus, the only option is to observe secondary products: parent molecules, and daughter molecules which result from their photodissociation, in the form of radicals, ions and atoms.

This task is complicated for the following reasons. Parent molecules produced directly from the nucleus are difficult to observe, because their spectral signatures (rotational and

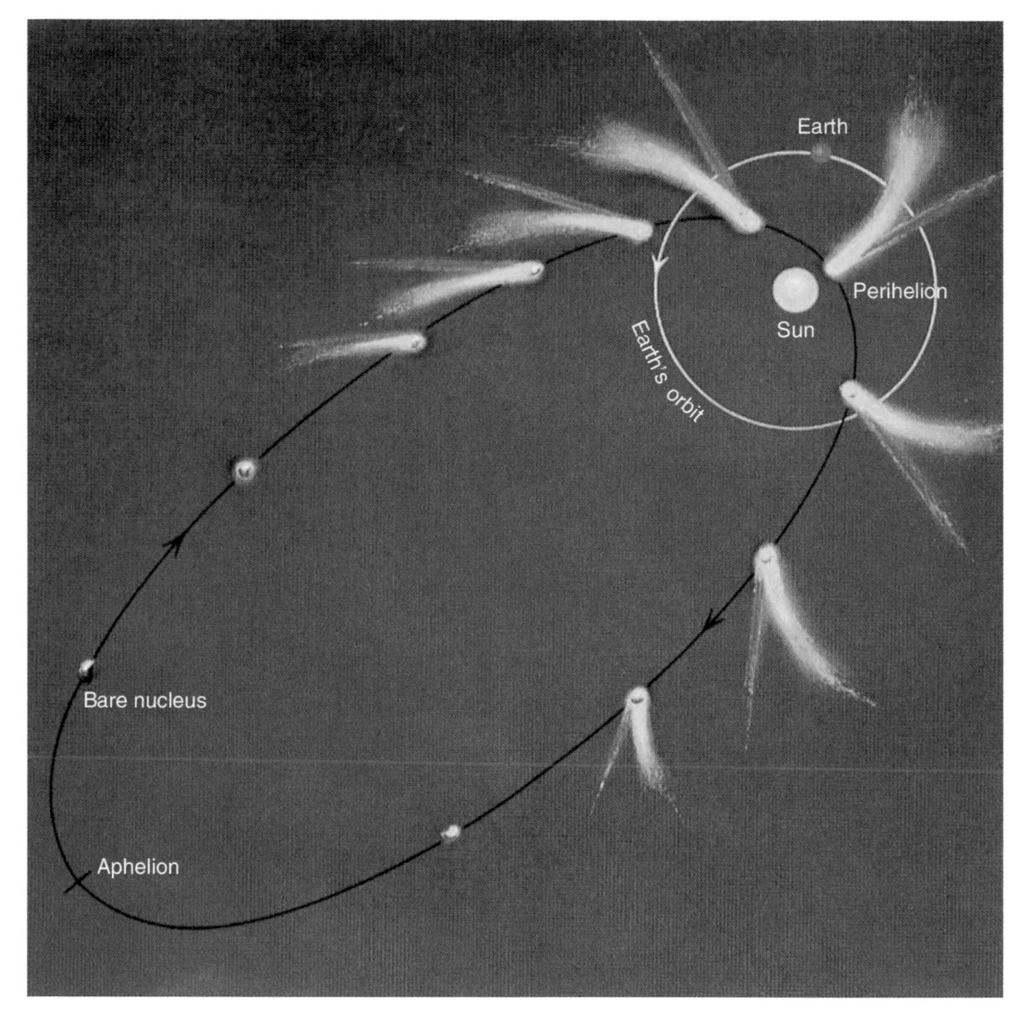

Figure 2.2.
Trajectory of a comet and evolution of its tails. Activity increases in the approach to perihelion, as do the sizes of its dust and plasma tails. These may reach lengths as great as several tens of millions of kilometres, whereas the nucleus never exceeds a few kilometres. Maximum activity occurs in the neighbourhood of the perihelion. The tail of ionised gases, always directed away from the Sun, is straight, in contrast to the dust tail, which is curved. Courtesy of Encyclopedia Universalis, *Le Grand Atlas de l'Astronomie*, 1985.

vibrational-rotational bands) lie in the infrared and millimetre ranges. These are more difficult to observe than visible wavelengths. By contrast, radicals, ions and atoms produced by photodissociation of parent molecules exhibit strong spectral signatures. These result from electronic transitions which lie in the visible region. Secondary products have been observed since the beginnings of spectroscopy over a century ago, and were soon identified (O, C, C_2, C_3, CH, CN, CS, CO_2^+, H_2O^+, and so on).

Many different secondary products have been observed, and the ionisation and dissociation reactions are numerous. In the majority of cases, these observations are insufficient to determine the composition of the corresponding parent

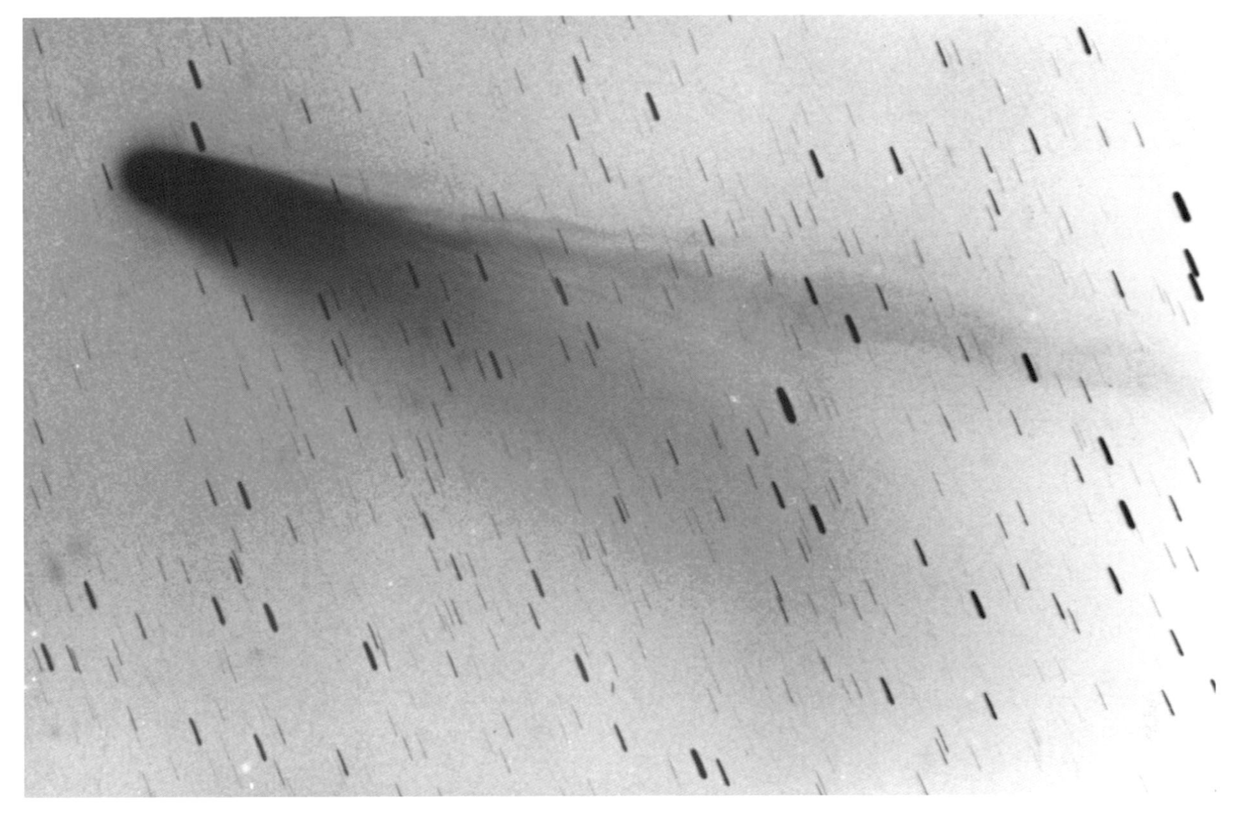

Figure 2.3.
Photo of an active comet with its two tails. Courtesy of IHW.

molecules unambiguously. Consequently, the best approach is to observe parent molecules directly in the infrared and millimetre regions, and ultimately to carry out *in situ* observation with space probes.

2.2 Cometary orbits

Comets are subject to the solar gravitational field and therefore satisfy Kepler's laws. Their trajectories are conic sections and can be described by their *orbital elements*.

A certain number of parameters are required to define such an orbit. Firstly, the *perihelion* is the point of the orbit which is closest to the Sun. The *nodes* are the two points of intersection between the comet's trajectory and the plane of the ecliptic (i.e. the plane of the Earth's orbit around the Sun), and the *ascending node* is the node at which the comet is crossing from the south side to the north side of the ecliptic. The *line of nodes* is the line joining the two nodes, which is also the intersection of the comet's orbital plane with the plane of the ecliptic. The *argument of the perihelion* is then the angle between the direction of the perihelion and the line of nodes. Finally, the *inclination* is the angle between the plane of the comet's orbit and the plane of the ecliptic. There are thus six orbital elements defining the orbit of a comet (see the box below).

Orbital elements of a comet

The orbital elements of a comet are as follows:

- the time T of perihelion passage;
- the distance q of the comet at perihelion;
- the argument of the perihelion ω;
- the longitude Ω of the ascending node;
- the inclination i of the orbital plane;
- the eccentricity e of the conic section.

In the case of an elliptical orbit, the following can be added:

- the semi-major axis a of the ellipse;
- the distance Q of the comet at aphelion (point on the orbit at maximal distance from the Sun);
- the period P of the orbit.

However, the last three items follow from the six fundamental elements given above, using the relations

$$q = a(1 - e),$$

and

$$Q = a(1 + e),$$

and the fact that the period P in years and the semi-major axis a in AU are related by Kepler's third law

$$a^3 = P^2.$$

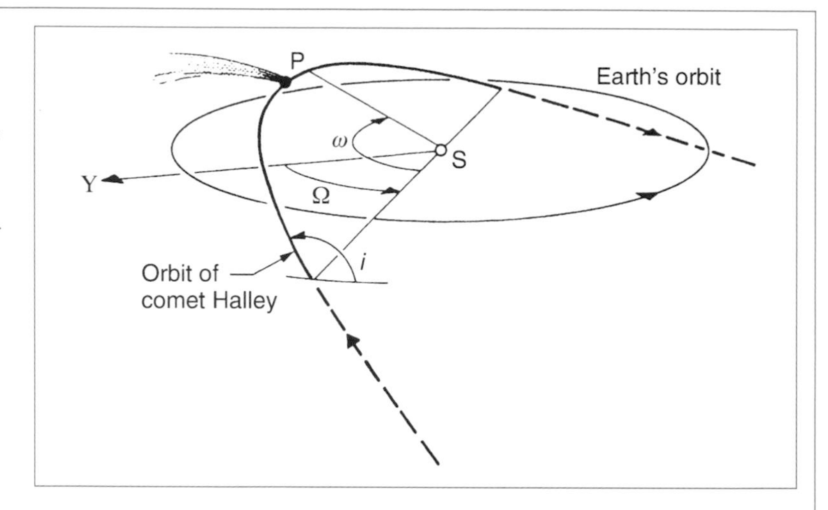

Figure 2.4.
Orbital elements of a comet (P/Halley).

Elliptical orbits correspond to eccentricities $e < 1$. The limiting case of a circular orbit is given by $e = 0$. With very few exceptions, cometary orbits are highly eccentric, whereas planets and most asteroids have quasi-circular orbits, with $e < 0.2$. Very long-period comets are highly eccentric, the value of e approaching unity. In most cases, their orbits cannot be distinguished from a parabola ($e = 1$). Eccentricities greater than unity correspond to hyperbolic orbits. Only 4% of known comets have hyperbolic orbits, but all these have eccentricities very close to unity, the most hyperbolic being Bowell 1982 I, for which $e = 1.057$. Needless to say, these comets escape from the Solar System. Whenever calculation has been possible, it transpires that these comets are very long-period comets whose orbits have been recently modified by planetary perturbations. A very modest increase in e (of the order of 10^{-4}) suffices to transform an elliptical orbit into a hyperbolic orbit. The case of a comet coming from outside the Solar System, whose eccentricity might therefore greatly exceed unity, has not yet been encountered.

As mentioned earlier, comets have a wide range of periods, for example, 3.3 years in the case of comet Encke, and 76 years for comet Halley. The period of comet Encke is the shortest known to date. Other comets, however, appear for the first time, and are called *new* or *non-periodic comets*. Their trajectories are quasi-parabolic. Although most comets remain unnoticed by the layman, being invisible to the naked eye, astronomers identify several new comets and several dozen periodic comets each year.

2.3 Cometary nomenclature

The nomenclature for comets is based on their observation from Earth. With the tremendous progress in astronomical instrumentation, which allows observation of fainter and fainter objects, the number of comets observed has risen dramatically. Up until 1994, the International Astronomical Union (IAU) adopted the following system of nomenclature. When a comet was observed in a first or a new passage, it was attributed a provisional name consisting of the year followed by a lower case letter determined chronologically in alphabetical order (1994a for the first discovery of 1994, then 1994b for the second, and so forth). For several years the whole alphabet was used up before the end of the year, and the sequence was continued a_1, a_2, and so on. The orbital details were calculated from observations, and at some time during the following two years, the comet received its permanent designation. The latter contains the year of perihelion passage followed by a Roman numeral, corresponding to the chronological order of its perihelion passage (1994 I, 1994 II, and so on). At each apparition, periodic comets thus received a new designation. Moreover, they carry the name or names of those who were first to observe them, or indeed in the past, the name of the first person to calculate their orbit (as in the case of Halley and Encke). Some comets are named after the observatory (e.g. Tsuchinsan) or the instrument (e.g. SOLWIND, IRAS, SOHO) used in their discovery. And some ancient or historic comets are referred to by those names that have become common usage, such as the Great Comet of 1910 and the Eclipse Comet. A list of some exceptional comets is given in the Appendix.

For periodic comets of period less than 200 years, the symbol P/ is inserted before the name of the discoverer. If several such comets have been discovered by the same person, they are numbered in order, examples being P/Tempel 1, P/Tempel 2 and P/Schwassmannn–Wachmann 1, 2 and 3. In the category of periodic comets are comets of Halley type (20 years $< P <$ 200 years), and comets of the Jupiter family ($P <$ 20 years), so named because their orbits are governed by Jupiter's gravitational field. The list of known comets is regularly updated in catalogues published by the Central Bureau for Astronomical Telegrams, currently directed by Brian Marsden (born 1937). The 1996 edition recorded 1470 cometary apparitions, among which 162 were of the Jupiter family and 23 of Halley type.

This nomenclature had several disadvantages, one being that it generated a different designation for each passage of any periodic comet, whereas some of these periodic comets, such as P/Encke, can be observed throughout their whole orbit. The International Astronomical Union proposed a new system of nomenclature, which was put into practice for comets discovered since 1995.[1] Comets are now only named as they are discovered, by the year, followed by a capital letter which indicates the two week period of the year, and finally a figure indicating the chronological order in which they were found during that two week period. Thus, C/1996 B2 (Hyakutake) was the second comet discovered in the second fortnight of January 1996. There will no longer be any later 'permanent' designation. However, the tradition of associating names of discoverers has been maintained. Short period comets now have only one designation, whatever apparition they may be making. They also receive a number giving the historical order in which their period was determined (hence, 1P/Halley and 2P/Encke). 137 short period comets are thus numbered as of June 1998.

2.4 Interaction of comets with the solar wind

Although it may be extremely tenuous, the medium in which comets orbit is not a total

[1] In the present book, we have kept old style designations for comets observed before 1995. The correspondence between old and new style designations may be found in the Appendix.

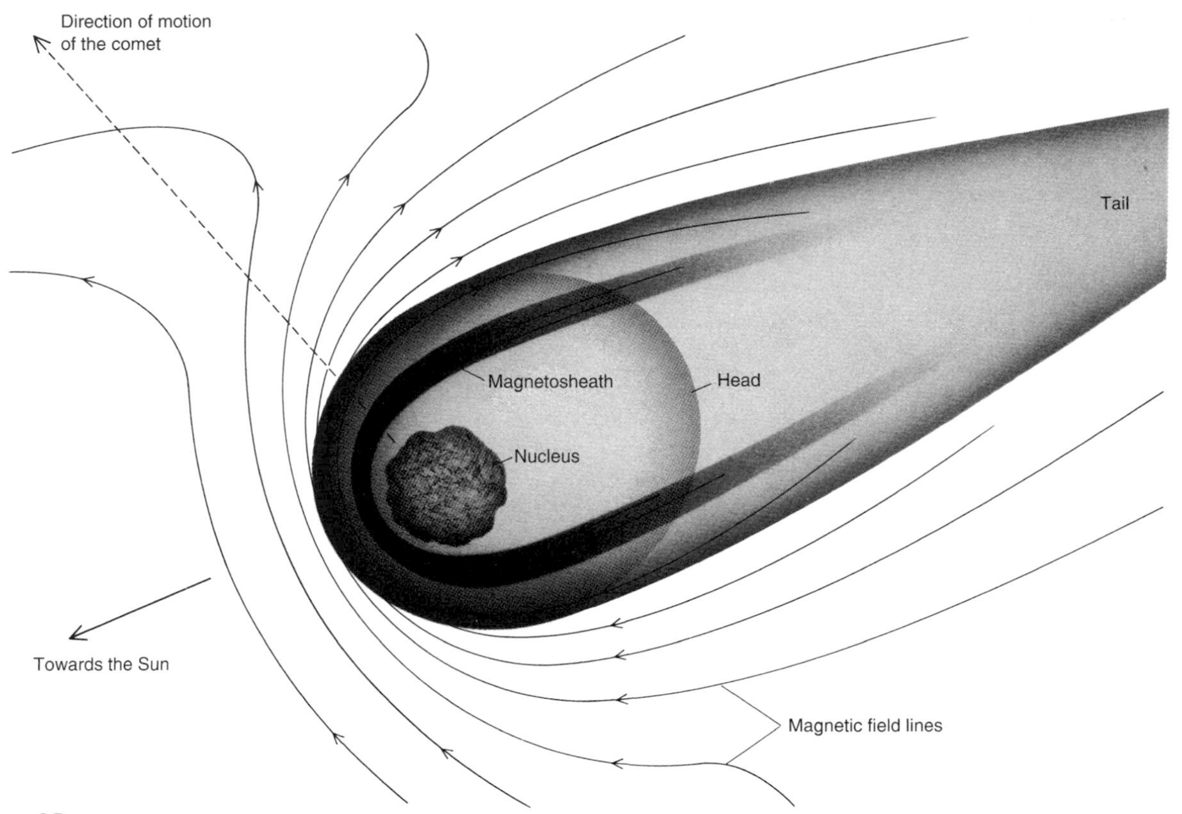

Figure 2.5.
Magnetic field lines of the solar wind are compressed when they enter the cometary atmosphere. The solar wind picks up cometary ions, and when it has penetrated deeply enough into the atmosphere, these pickup ions become sufficiently numerous to exert a pressure which can balance the pressure of ions liberated from the cometary nucleus. The lines of force of the solar wind can penetrate no further into the cometary atmosphere, and a region of zero magnetic field is thus conserved close to the nucleus. From Brandt and Nieder, 1986 [33].

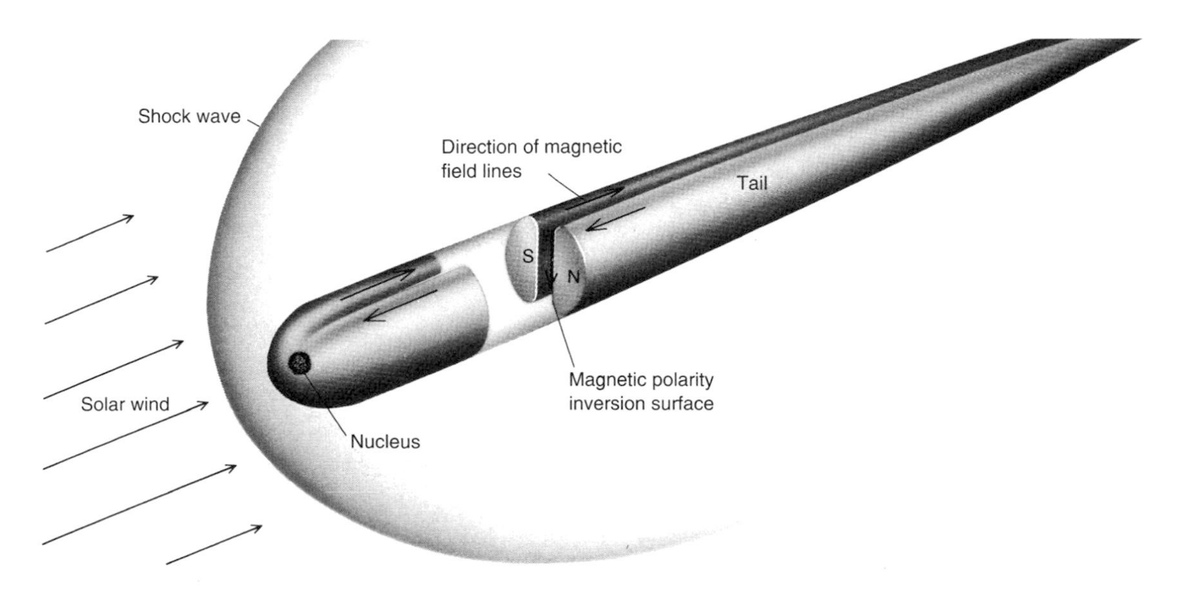

Figure 2.6.
Ions in the tail of the comet are channelled along magnetic field lines of the solar wind as it wraps itself around the comet. A *neutral sheet* separates lobes of opposite polarity of the magnetic field. When the supersonic solar wind has picked up enough cometary ions, it slows down suddenly, crossing a shock wave analogous to the *bow shock* caused by the passage of a boat through water. From Brandt and Nieder, 1986 [33].

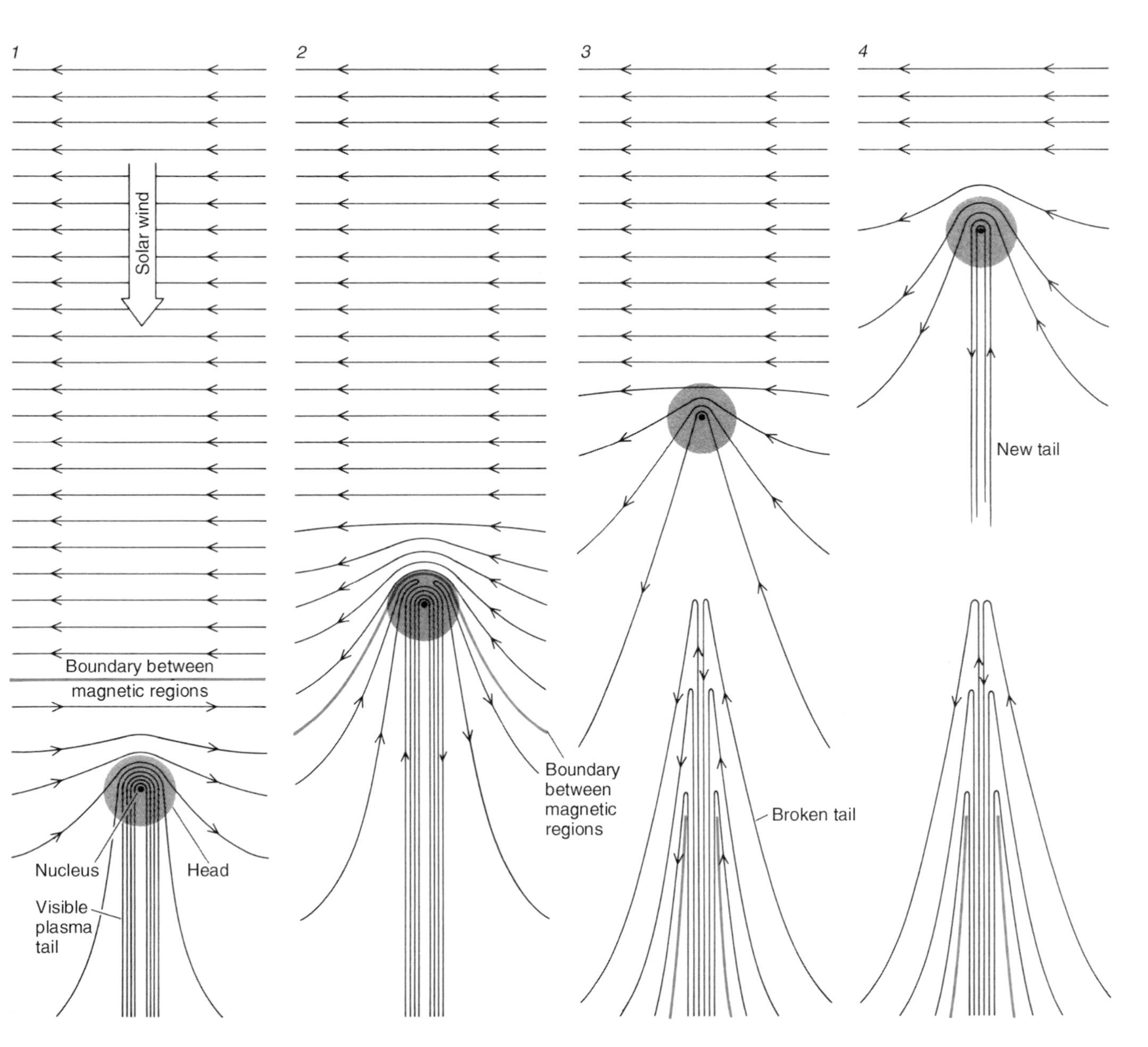

Figure 2.7.
The plasma tail disconnects when the comet moves from a region in which its magnetic field has the same polarity as that region to one in which it has the opposite polarity. The old field lines in the tail are thus broken through a process referred to as *magnetic reconnection*, retaining the ions trapped previously, but no longer linking up with the comet. The tail appears to separate from the comet (cf. Figs. 2.8 and 2.9) and a new tail is immediately formed, of the same polarity as the new magnetic field region. From Brandt and Nieder, 1986 [33].

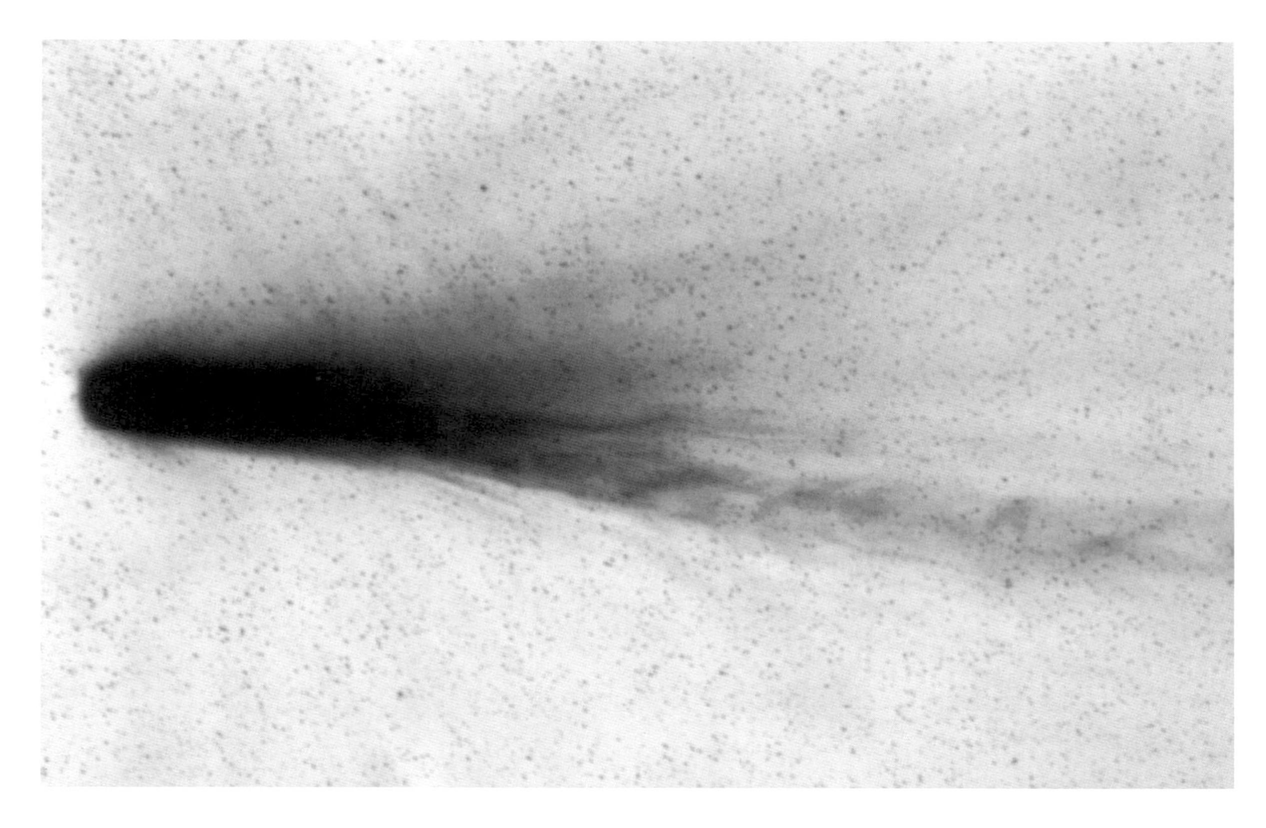

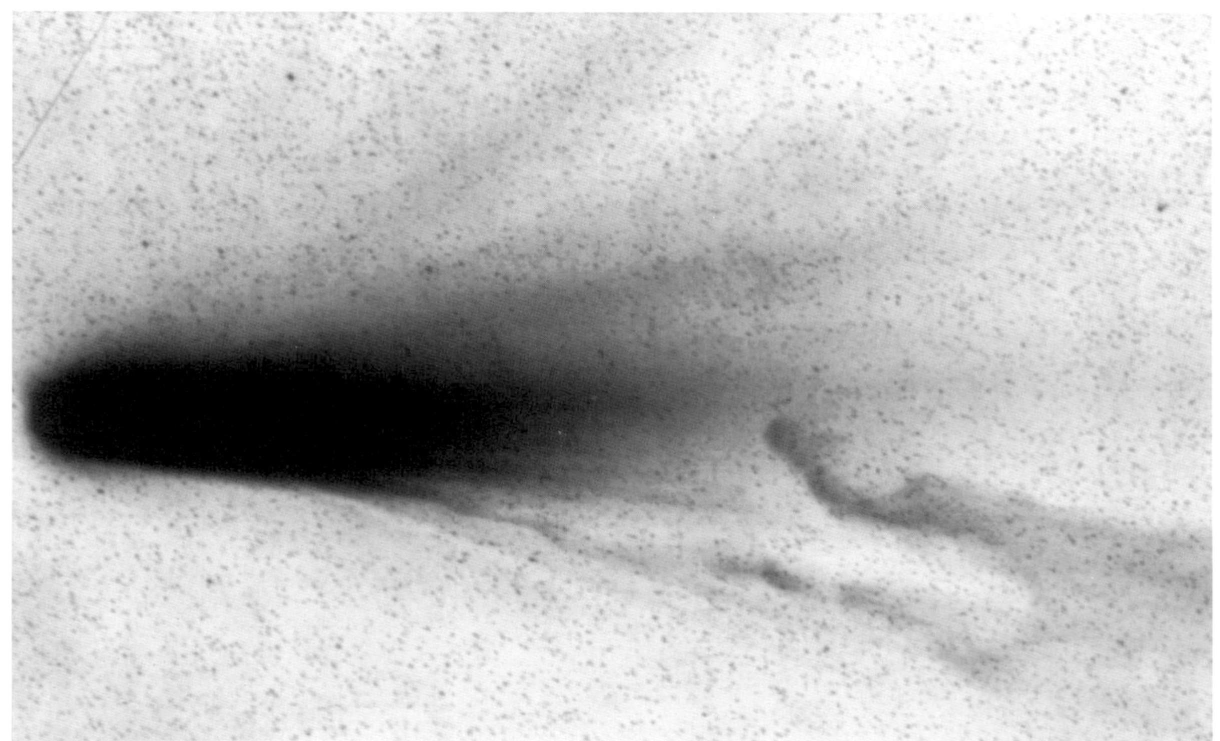

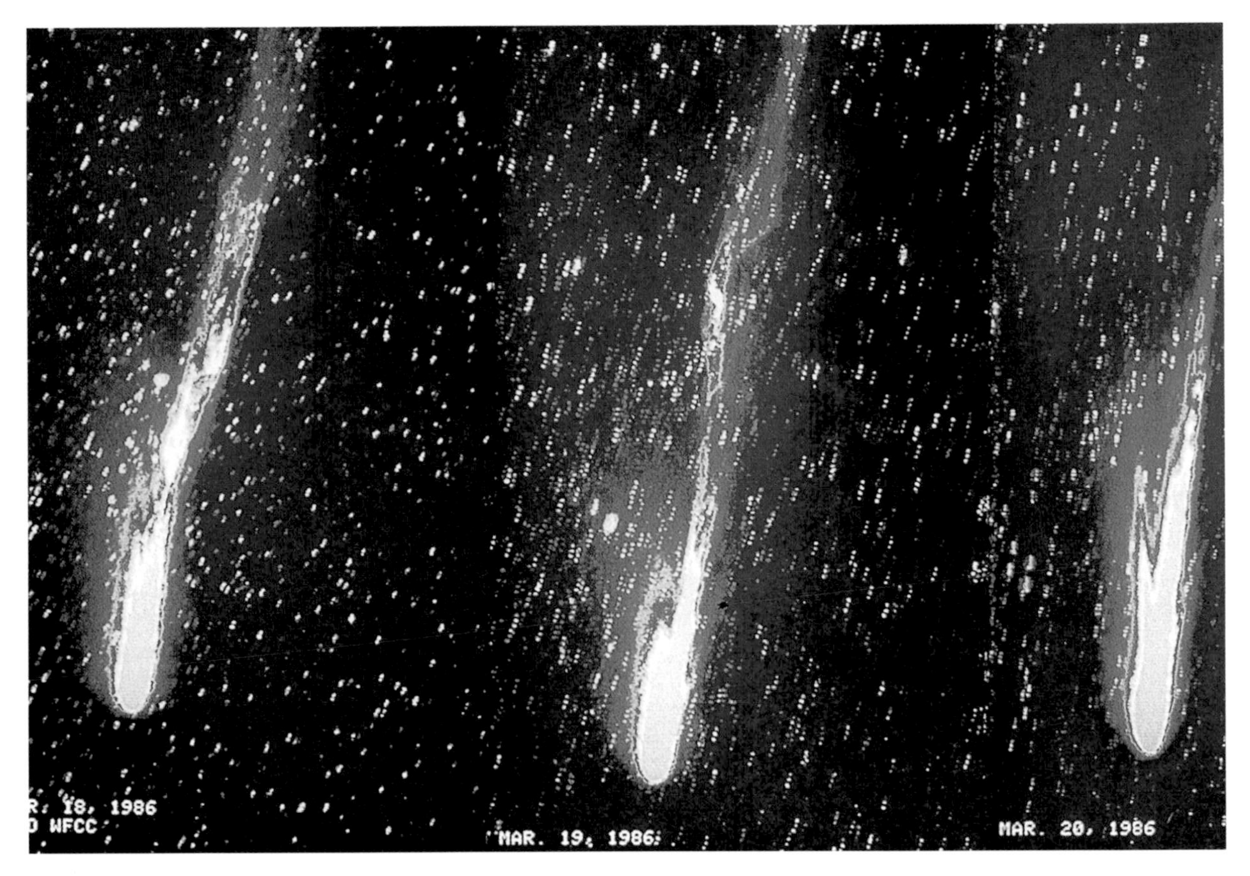

Figure 2.9.
These false colour images show evolution of the comet Halley plasma tail between 18 and 20 March 1986, under the influence of the solar wind. The images were made with a wide angle camera (100 mm objective) equipped with CCDs and a filter isolating CO^+ ion emission. Courtesy of ESO.

vacuum. The interplanetary medium is permanently subjected to a flux of particles from the Sun, called the solar wind, which was only discovered relatively recently. Indeed, it was in the 1950s that the German scientist Ludwig Biermann (1907–1986) found evidence for a continuous flow of ionised particles from the Sun, by observing the systematic action of this flow on ion tails of comets. A few years later, a theory was propounded by E. Parker (born 1927) which described the flow of this hot plasma, directly related to the solar corona and its high thermal conductivity. The temperature of the solar corona (of the order of one million degrees) is such that a stable stationary state requires its continuous expansion, in a monokinetic 'wind'. The solar wind is thus a high temperature plasma (that is, an ensemble of ionised particles, ions and electrons), which is in consequence highly conductive. Its ion composition is approximately 95% protons and 5% helium nuclei. It carries with it an interplane-

Figure 2.8. opposite
Photos of comet Halley taken on 9 and 10 March 1986, with the ESO's 1 m Schmidt telescope. The spectacular development observed on 10 March is due to the crossing of a boundary between two oppositely polarised magnetic regions of the interplanetary field, causing the plasma tail to disconnect. Courtesy of ESO.

tary magnetic field from the Sun. At a heliocentric distance of 1 AU, the speed of the solar wind is around 400 km/s, and its mean density about 5 ions/cm^3. The ion density decreases as the inverse square of the heliocentric distance. In periods of high solar activity, the density can increase by as much as a factor of ten. The magnetic field carried by the solar wind typically has strength between 1 and 10 nT. This too may increase by a factor of ten during large solar flares, at the origin of solar cosmic radiation.

How does a comet interact with the solar wind? Comets are too small to possess their own magnetic field, but their presence in the interplanetary medium causes a local modification in the magnetic field carried by the solar wind. While a comet is very distant from the Sun, there is almost no production of ionised particles and hence very little modification of the surrounding solar wind. However, when the comet becomes active, it produces an envelope of hydrogen atoms whose speed is negligible compared with that of the solar wind. Interaction begins when the solar wind penetrates this envelope. A shock wave is thereby created, somewhere between 100 000 km and 1 000 000 km upstream of the nucleus. Charged particles cannot move in the transverse direction relative to the magnetic field, but instead follow spiral paths along its lines of force. Cometary ions are thus trapped along the force lines of the solar wind. The latter gains in mass and is gradually slowed down, until it reaches a contact surface in the region where solar wind pressure is balanced by dynamic pressure from the coma.

Towards the nucleus, magnetic field lines are compressed, thereby increasing the magnetic field. They then fold around each side of the nucleus to form two lobes of opposite polarity. Between the two lobes, the magnetic field must pass through a value close to zero, and a current sheet of extremely high plasma density should be observed. This scenario, due to Ludwig Biermann and the Swedish scientist Hannes Alfvén (1908–1995), has been borne out by space exploration of comets.

The structure of these straight cometary tails has intrigued observers for a long time. They could be analysed at leisure from repeated observations using wide field photographic plates. The ion tail is often characterised by a series of symmetric pairs of ion beams, of diameter between 1000 and 10 000 km, tilted at a small angle to the central axis of the tail. Using magnetohydrodynamic models, it has been suggested that the existence of these ion beams is correlated with the presence of tangential discontinuities in the interplanetary magnetic field, and that the formation of an ion beam might be due to narrow filaments of high density but low field, separated by regions having opposite characteristics. The final stage in the development of these ion beams is their disappearance into the central ion tail.

Apart from ion beams, cometary plasma tails exhibit a wide range of morphological features, including waves, nodes and spirals, which vary both in intensity and in spatial position, over time scales of several hours. Another spectacular phenomenon characteristic of plasma tails, which has frequently been observed and known for many years, is an apparent disconnection of the plasma tail at the back of the coma, followed by formation of a new tail from ion beams converging towards it. This phenomenon is attributed to the passage of the cometary nucleus through a boundary in the magnetic field of the solar wind, between regions of opposite polarity. A new plasma tail is then observed, whilst the old tail, of opposite polarity, disconnects.

2.5 Cometary origins

Although we may now know the trajectories of the comets, it is not yet understood where they come from. Did they form in their present locations, and if not, what forces have brought them there?

Once again, it is the science of celestial mechanics which gives the beginnings of an answer. Indeed, just as the equations of mechanics can predict the future motion of comets, provided planetary perturbations are taken into account, they can also retrodict these paths, by including perturbations due to the giant planets, to give the original orbits of these objects. Such was the task undertaken by the Dutch astronomer Jan Hendrick Oort (1900–1992), towards the middle of the twentieth century. Based on a study of about twenty periodic comets of well-known trajectory, he was able to demonstrate that the original orbits of these objects, before the influence of planetary perturbations, were without exception quasi-parabolic. In other words, not one of them originated either from outside the Solar System or from the inner Solar System. In these cases their orbits would have been either hyperbolic or elliptical. The comets he studied all seemed to originate in a region situated somewhere between 40 000 and 100 000 AU from the Sun, at the very limit of the Solar System. This great reservoir was called the *Oort cloud*. Subsequent work, led by Brian Marsden on a larger sample of comets, has only confirmed the first result. The Oort cloud may contain as many as 10^{12} comets. Under the effect of perturbation by nearby stars, 5 to 10% of the cometary population could be ejected from the cloud, and when a new comet comes close enough to the Sun to suffer planetary perturbations, it may find itself trapped in a short period orbit. The comets we observe therefore constitute only a very small fraction of the whole group of these celestial bodies.

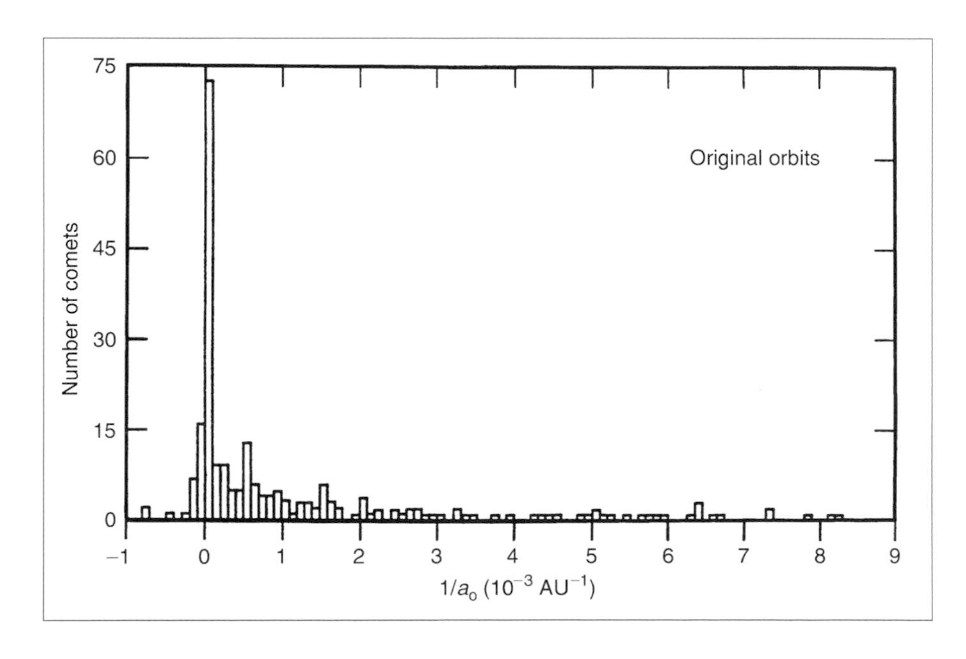

Figure 2.10. Distribution of periodic comets as a function of the reciprocal semi-major axis of the ellipse which constitutes their orbit, $1/a_0$. Such a diagram led to Oort's hypothesis of a cometary reservoir. After Mumma *et al.*, 1993 [9].

It is clear then that comets belong to the Solar System, although the reservoir which contains them is so remote (its heliocentric distance situates it somewhere between a quarter and a half of the way to the nearest star) that it must be identified as part of the local interstellar medium. Such a result could only enhance astronomers' interest in comets. Evolving in a medium in every way comparable to the local interstellar medium, and protected from any later modification, comets would seem to be perfect witnesses to the physicochemical processes set up during formation of the Solar System.

But one difficulty remains to be faced. How did the comets come into being at several tens of thousands of AU from the Sun, in a medium of such very low density? Even today, this problem seems insoluble, and a different scenario is gaining some support: comets may have formed, together with a vast quantity of other small bodies of various sizes, at more than 5 AU from the Sun, somewhere beyond the orbits of Jupiter and Saturn. Perturbations due to the Jovian planets – Jupiter, Saturn, Uranus and Neptune – would then have ejected a certain number of these objects into the Oort cloud, from whence a small percentage occasionally return (see box entitled *Dynamical evolution of comets*).

If this scenario proves to be accurate, many fossil comets should still be present where they formed, beyond the orbit of Neptune, in a region free from the perturbations of Jovian planets. This region constitutes the so-called *Edgeworth–Kuiper belt*, named after the free-lance Irish astronomer Kenneth E. Edgeworth (1880–1972) and the American astronomer Gerard Kuiper (1905–1973), who were the first to put forward the idea. A mere hypothesis a few years ago, the existence of trans-Neptunian objects has now been confirmed, thanks to recent instrumental progress, together with the construction of very large telescopes. It will be seen in Chapter 10 that many of these bodies, either asteroids or comets, have been detected recently. This line of research should see a considerable development over the next few years.

It is a small step from the outer limits of the Solar System to the interstellar medium. Whatever their exact origins, it is clear that comets have formed in a cold and low density medium from some fragment of the outer primordial nebula. It is reasonable to expect certain similarities between the chemical composition of comets and that of the interstellar medium. It will be seen in Chapter 9 that parent molecules produced from cometary nuclei have also been observed in the interstellar medium. Moreover, organic compounds which seem to be present in cometary dusts, as on the nuclear surface of comet Halley, might well resemble refractory organic compounds observed in the interstellar medium. The latter could have been produced through irradiation of organic ices by intense ultraviolet radiation or a flux of high energy particles.

The scope of these analogies between cometary matter and interstellar matter opens a new door for cometary science. From time to time, comets come close enough to the Sun and Earth to deliver up their secrets. If they are even partially representative of conditions prevailing in the interstellar medium, they constitute for us an invaluable tool for study of star formation. They teach us not only about the origins of the Solar System, but also about the physicochemistry of the interstellar medium and mechanisms of stellar formation and evolution. As such they constitute for astronomy a genuine 'window' looking out over the Universe.

Dynamical evolution of comets

Regular apparitions, on a historical scale, of periodic comets should not make us forget that this phenomenon is only a temporary stage, on an astronomical scale, in the evolution of a celestial body. Significant loss of matter at each perihelion passage of the comet implies an inescapable reduction in cometary activity. In order to explain the existence of currently active comets, this phenomenon must be balanced by a continual injection of new comets. Such is made possible by dynamical evolution of cometary orbits.

A pictorial representation of relations between various families of comets, borrowed from the Slovakian astronomer Lubor Kresàk (1927–1994), is given in Fig. 2.11. The *Oort cloud* groups together those comets whose periods exceed several million years. It should be emphasised that no member of this cloud has been directly observed, owing to its remoteness. The Oort cloud feeds the category of *new comets*, which are highly eccentric; their aphelia remain in the cloud, whilst their perihelia are in the inner Solar System, which renders them observable. These new comets have periods of the order of 10^3 years and orbits quite indistinguishable from a parabola. Following planetary perturbations of a comet, either its orbit becomes hyperbolic, and it leaves the Solar System for ever, or its eccentricity decreases, and it evolves towards a short period orbit of *Halley type*.

This schematic view is quite likely incomplete, considering that only those comets whose perihelion distance is less than a few AU are actually observable, and that the most distant objects known in the Solar System lie within a heliocentric distance $r_h < 45$ AU. The Oort cloud, assumed to have a spherical distribution, can thus explain new and long-period comets, whose orbital planes are randomly tilted relative to the ecliptic. But comets in the *Jupiter family* have orbital planes lying close to the ecliptic, like planets and asteroids. It is difficult to imagine how comets originating in the Oort cloud could have preferentially acquired such orbits. This was the problem which inspired Kuiper to propose a reservoir of comets forming a ring within the ecliptic itself, known as the *Edgeworth–Kuiper belt*. The inner region of this belt, within range of planetary perturbations, would feed the group of Jupiter family comets. As we shall see in Chapter 10, the first objects of this belt are just beginning to be discovered.

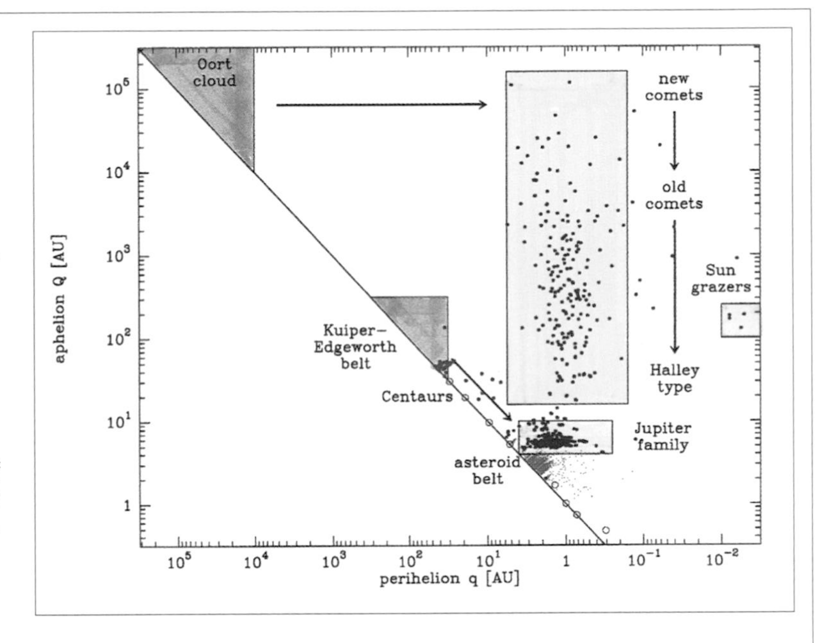

Figure 2.11.
q–Q diagram and evolution of cometary orbits. Objects in the Solar System are represented by coordinates q (perihelion distance) and Q (aphelion distance). The planets (*red circles*) have almost circular orbits and hence lie close to the straight line $Q = q$. Asteroids (shown as a cloud of *tiny blue dots*) are essentially confined to the region of the ecliptic known as the asteroid belt ($2 < q, Q < 3.2$ AU). All recorded comets of elliptical orbit (for which Q is known) are represented by *red dots*. Objects of the newly discovered Edgeworth–Kuiper belt and Centaurs are represented by *green dots*. Zones corresponding to different families of comets have been identified on the diagram, and *arrows* show presumed relations between them, resulting from evolution of cometary orbits. From Kresàk, 1994 [59].

3 Visual observations

3.1 In search of comets

A comet appears to observers as a fuzzy patch, moving relative to background stars. Whilst searching for comets, it is important not to confuse this nebulosity with that of galaxies, galactic nebulas or unresolved star clusters. Astronomers need catalogues and sky charts, and it was with this aim in mind that Charles Messier (1730–1817) made the first catalogue of nebulous objects (now known by his name), at the Paris observatory in the eighteenth century. Motion across the sky, a characteristic used in photographic comet searches, is the determining criterion, and the same searches also detect asteroids.

Among the twenty or thirty comets observed each year, more than half are previously recorded periodic comets whose return has been predicted. Such comets are sought in the neighbourhood of positions predicted by celestial mechanics, and observations often use powerful telescopes. Generally speaking, they provide few surprises. There may be slight deviations relative to expected positions, due to non-gravitational forces; and occasionally a comet does not reappear, due to extinction or inactivity, breaking up of the nucleus, inadequate knowledge of orbits, or some other cause.

Altogether more interesting are unexpected comets. These may be discovered by any of several different procedures. Indeed, there is no truly systematic search programme by professional astronomers, bearing in mind that it would take several years to cover the whole sky with any acceptable sensitivity. However, many comets are discovered accidentally during systematic searches for asteroids and supernovas. Many other chance discoveries are made whilst analysing photographic plates for quite different reasons.

As we shall see later, comets have also been found by satellites, for example, during the systematic sky survey by the infrared satellite IRAS (Chapter 6), and by coronagraphs on board satellites SMM, SOLWIND and SOHO (Chapter 7) as they approached the Sun.

In addition, it should be emphasised that amateurs have always played a crucial role in discovering comets. At the present time, a small number of well-trained comet hunters, with detailed knowledge of the sky and only simple observing instruments, devote a great deal of time to these searches. Indeed, a small telescope or large binoculars, which are easy to handle and have great light-gathering power, are good enough to detect comets down to magnitude ten. It is notable that the same names regularly appear as comet discoverers in cometary nomenclature.

Some comet hunters have achieved a certain fame, including Charles Messier and Pierre

Méchain (1744–1804) in the nineteenth century, then Jean-Louis Pons (1761–1831), who started out as a doorman at the Marseilles observatory and discovered twenty-eight comets between 1801 and 1827.

Among the most active comet discoverers of our own day are the professionals Carolyn and Eugene Shoemaker (1928–1997), with twenty-eight comets to their name, found with the Mount Palomar Schmidt telescope; the Australian amateur William Bradfield, who has found thirteen comets in the Southern Hemisphere; and David Levy, who has found ten comets with the Shoemaker team, and seven more as an amateur. Teams of Japanese amateurs have also been very active.

3.2 Astrometry and determination of orbits

One of the preoccupations of cometology has always been to measure cometary positions accurately enough to determine orbits. The aim is partly to predict the evolution of new comets and foresee the return of periodic comets, and partly to gain more accurate knowledge of long term orbital evolution. This leads to a better understanding of non-gravitational forces affecting cometary nuclei. Note also that observers require very accurate ephemerides for certain types of observation, such as radio observations, which are carried out 'blind', as it were. The same is true when space missions are set up.

Position measurements are made on photographic plates or digitised CCD images, by comparison with reference stars. A precision of 0.2 arcsec can be attained. When a new comet is discovered, its orbit can be reliably estimated after only a few days of measurements. In theory, only three astrometric measurements are required, provided they are sufficiently widely separated in time. At this stage, the eccentricity of the orbit cannot be measured, this becoming possible only after several weeks or months of observation. What is found then is an osculating parabola. In the following step, a set of orbital elements can be fitted by the least squares method. This can reproduce the whole astrometric data set with great accuracy, often to within 0.5 arcsec. Of course, it is now no longer enough to make a two-body calculation of the orbit. Perturbations due to all the planets must be taken into account, Jupiter's influence being the major factor. Gravity alone is not always sufficient either, when modelling a comet's motion in all its detail. Reaction forces due to ejection of gas and dust during active periods must be included. These are the so-called non-gravitational forces, which will be discussed further in Chapter 7.

3.3 Evolution of visual magnitudes

The general features of comets are known. Their luminosity is explained by the scattering of solar light from the nucleus and from dust. On top of this are superposed molecular frequency bands of fluorescing molecules in the gaseous atmosphere. For an active comet, the brightness of the nucleus is totally negligible compared with that of gas and dust in its atmosphere. A comet's brightness therefore directly indicates the rate at which matter is being ejected.

However, this 'brightness' is a quantity which is difficult either to define or to measure, for the exact extent of the body is hard to determine, and it is only weakly contrasted with the sky background, especially in twilight at dawn or dusk, or when the Moon is out. On the other hand, brightness estimates are often the only data available concerning historic comet activity; and for more recent comets, it is the most

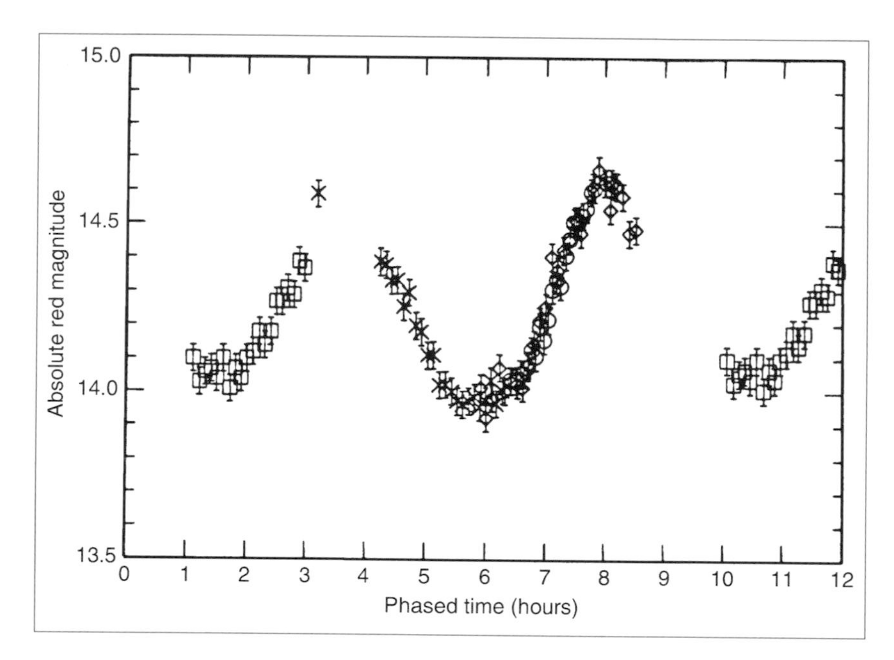

Figure 3.1.
Light curve observed for the comet P/Tempel 2, showing a period of 8.95 hours due to rotation of the nucleus. The amplitude of these brightness variations, around 0.7 magnitudes, is explained by irregularity of the nucleus. From Jewitt and Luu, 1989 [54].

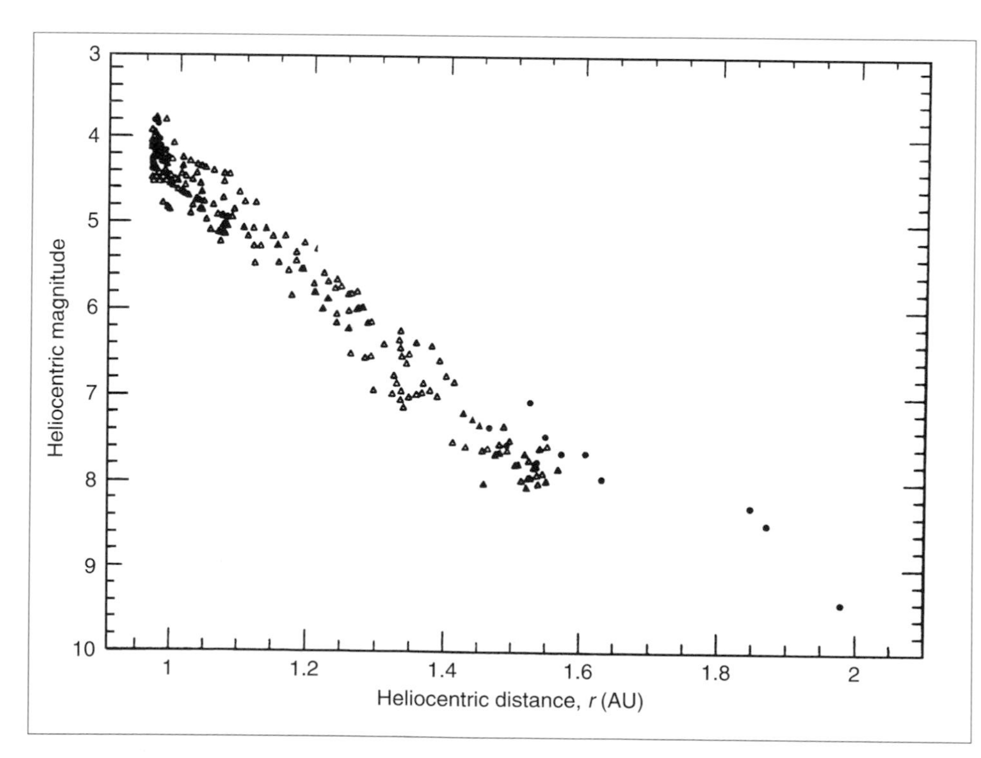

Figure 3.2.
The heliocentric magnitude of comet P/Swift–Tuttle as a function of heliocentric distance (*triangles* for observations before perihelion and *full circles* for those after perihelion). Measurements come for the main part from observations by amateur astronomers. Courtesy of the *International Comet Quarterly*, January 1994.

common form of information, thanks to the availability and enthusiasm of amateur astronomers. Today, such information is gathered and entered into data bases which group together tens of thousands of observations.

Most observations assess the *total visual magnitude*. In most cases, the detector used is the human eye, or else a photometer with a filter reproducing the spectral response of the eye. In order to make measurements more uniform, standardised observation procedures have been defined. These consist in comparing the brightness of the comet with the brightness of standard reference stars viewed out of focus.

Attempts have been made to model long term evolution in the visual magnitude m_v of comets as a function of their distance from the Sun and the Earth, using simple laws (see box entitled *Cometary magnitudes*).

In reality, comets often display quite different behaviour to that predicted by such simple models. This fact demonstrates the diversity and complexity of processes governing cometary activity. These may involve the 'ignition' of activity below certain heliocentric distances, and its extinction or weakening close to the Sun for certain new comets; or alternatively, late activation due to delay (hysteresis) in heating of the nucleus for short period comets; and varying behaviour before and after perihelion, revealing a seasonal phenomenon for periodic comets. Many other types of process may also occur.

When a new comet appears, it has become common practice to predict its evolution by applying the simple law described in the box with index $n = 4$, in order to decide whether or not to organise an observation campaign. On many occasions, such predictions have proved misleading and observations disappointing. This is exemplified by comet Kohoutek 1973 XII, discovered at the beginning of 1972. Comet Kohoutek never fulfilled the spectacular expectations inspired by brightness predictions for its perihelion passage a year later.

Far from the Sun, if a comet is inactive, its magnitude is entirely due to its nuclear brightness. A detailed photometric study of short term changes in nuclear magnitude may reveal a periodic variation in the *light curve*, indicating the rotation of an irregular body. This method has

Cometary magnitudes

Cometary radiation flux Φ, like stellar radiation flux, is given in terms of a scale of magnitudes. This scale is logarithmic, the magnitude m increasing by 2.5 units when the intensity decreases by a factor of 10:

$$m = -2.5 \log \Phi + \text{constant}.$$

The zero on this scale is defined conventionally by fixing the star Vega (α Lyræ) to zero. The magnitude of the faintest objects that can be distinguished with the naked eye is about 5. The long term evolution of the visual magnitude m_v of a comet, as a function of its distances Δ from the Earth and r_h from the Sun, can be represented by a law of the kind

$$m_v = H_0 + 2.5k \log +2.5n \log r_h,$$

where H_0 is the *absolute magnitude* of the comet, and k and n are indexes characterising the law of brightness variation. k is close to 2, corresponding to the ideal case of a $1/\Delta^2$ variation in brightness, although deviations from this are observed for comets close to Earth, whose dimensions are difficult to perceive, and whose brightness is systematically underestimated. Brightness curves are usually converted to a standard distance $\Delta = 1$ AU in order to compare one comet with another. Typical values of n lie between 3 and 4. Far from the Sun, before activity begins (so that there is no coma and the body resembles a star), the brightness variation is just that of a bare nucleus. It then obeys precisely the same law as do the asteroids, with $n = 2$:

$$m_v = H_n + 5\log\Delta + 5\log r_h.$$

The *absolute nuclear magnitude* H_n is proportional to the exposed surface area of the nucleus. If the albedo of the nuclear surface is known, the mean radius of the nucleus can be estimated.

made it possible to study the rotation of several hundred asteroids. Measurements are more difficult for cometary nuclei, for two reasons. Firstly, they are smaller bodies and secondly, they are further from the Sun and the Earth at the time of study, because they must be observed before activity begins. Such studies have thus been successful for only a few objects to date (see Table 7.1). The rotational periods found, lying between 6 hours and 3 days, closely resemble those of asteroids and larger planets. It is a remarkable fact that, in the Solar System, small bodies do not rotate faster than larger ones, something which might be expected if angular momentum were uniformly distributed. This may result from some process which is more efficient in slowing down rotation in smaller bodies.

With the exception of Chiron, the mean radii of cometary nuclei accessible to photometric measurement lie between 1 and 20 km. The amplitude of light curve variations can be used to assess irregularities in the nucleus. For a given object, the ratio of maximum and minimum dimensions can reach a value of two; such is the case for comet Halley. Among the smaller bodies of the Solar System (small natural satellites, asteroids and comets), an increase in irregularity is observed as their size decreases. Indeed the accretion process which formed them, and collisions with outside bodies, both tend to produce spherical shapes once a certain size is exceeded. Large bodies which were heated up during their formation, to such an extent that they may have become fluid, can also acquire a spherical shape under the effect of their own gravitational field.

One comet can be compared with another through its absolute magnitude H_0, as explained in the box entitled *Cometary magnitudes*. Some values are given for exceptional comets in the Appendix, Table A.1. Figure 3.3 sums up what is known about the statistical distribution of these magnitudes for short and long period comets. This statistical study is complicated by two factors. Firstly, only those comets with a specific history can be compared. Furthermore, old comets must be included in order to obtain a large enough sample, and these measurements are difficult to calibrate.

3.4 Cometary imaging

We saw in Chapter 1 that the spectacular nature of comets has always made them a choice subject for popular pictorial representation. Recording and transcription of images is essential for cometary scientists to be able to study the morphology and evolution of these objects.

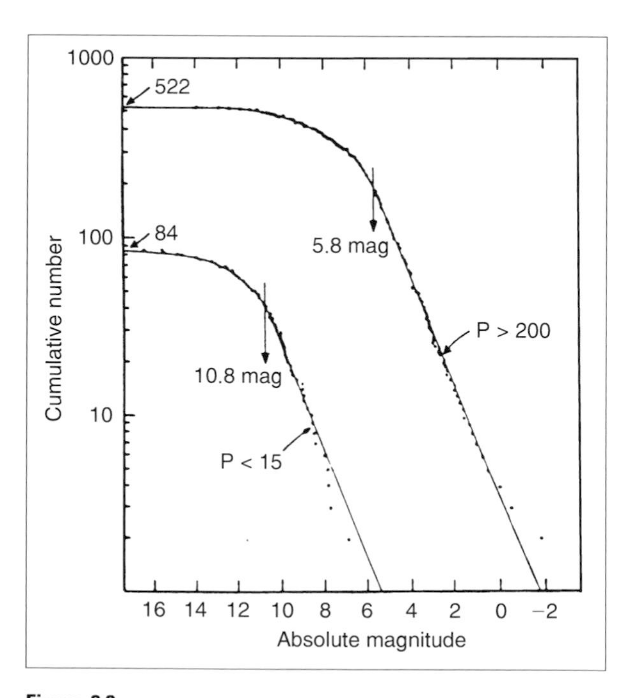

Figure 3.3.
Distribution of absolute magnitudes for comets. The curves show the number of comets brighter than absolute magnitude H_{10}. The two curves correspond to comets with periods $P < 15$ and $P > 200$ years. Flattening of curves at higher magnitudes is due to inadequate sampling of fainter comets. From Hughes, 1987 [52].

Figure 3.4.
The Pic-du-Midi Observatory (France) at an altitude of 2860 m. The largest dome houses a 2 m telescope. The site, which is often free of atmospheric turbulence, is one of the best in the world for planetary and cometary imaging observations. Courtesy of Pic-du-Midi Observatory.

Drawings have been used from the very beginning. Photography represented a genuine revolution, providing an objective, faithful and sensitive record, by means of long exposures, but drawings are nevertheless still used particularly by amateur observers. Indeed, visual observations can discern faintly contrasting details, such as dust jets, which are hard to distinguish on photographic plates. Modern observation uses CCD (or *charge coupled device*) cameras which are more sensitive and possess a far greater dynamic range than ordinary photography. With a photographic plate, we must make a series of exposures of different length in order to study both faint and bright details without the risk of saturation, whereas a single CCD exposure suffices. This technique is now relatively cheap and easy to use, and has become available to amateurs. It produces digitised images which can be directly processed by computer, hence combining the other great revolution in this area. Combined use of CCDs and image processing software has thus become an ideal tool for cometary imaging, where faintly contrasting details

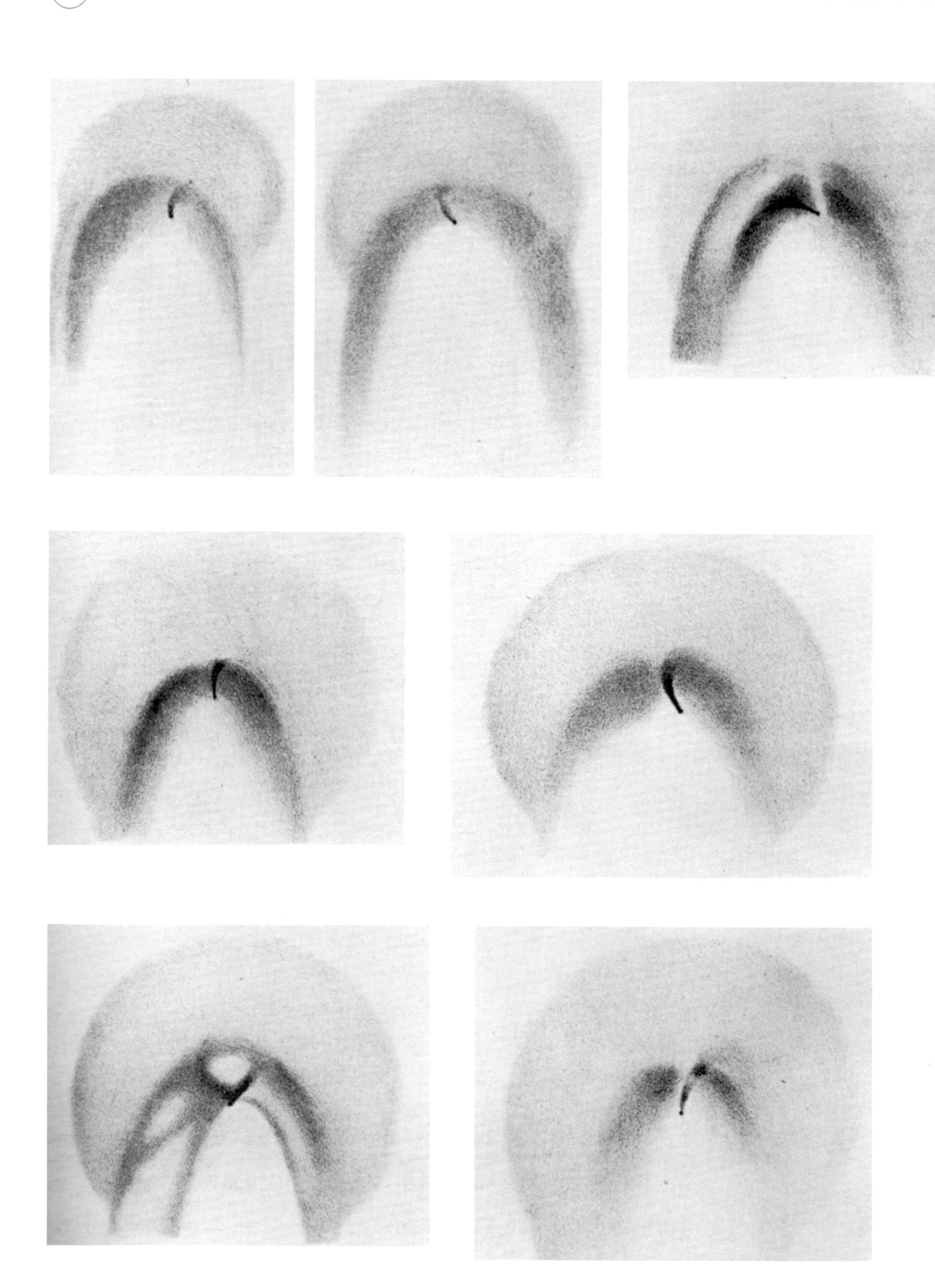

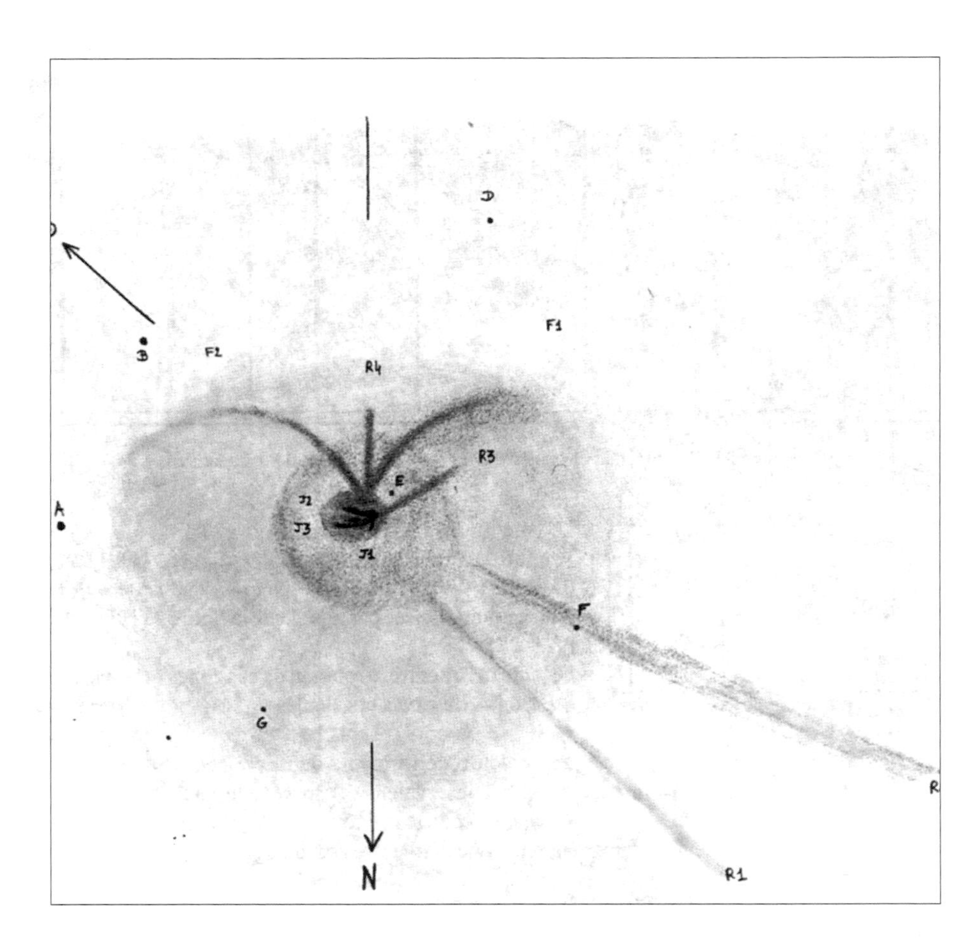

Figure 3.6.
Dust jets from comet P/Swift–Tuttle. Drawings made by an amateur astronomer on 21 November 1992, when the comet passed at 1.2 AU from Earth. Courtesy of S. Garro.

must be discerned on an image that originally contained a wide range of brightnesses.

For comets, there are two types of imaging, large scale and small scale.

Large scale imaging refers to fields of view lying between a fraction of a degree and several degrees, which require wide field instruments such as the Schmidt telescope. This is aimed at study of cometary tails (see the box entitled *Dynamics of dust tails*).

Figure 3.5. opposite
Dust jets from comet P/Swift–Tuttle. Drawings made by visual observation using a small telescope, during its 1862 passage at 0.35 AU from Earth. J.F. Julius Schmidt, Athens Observatory.

Small scale imaging makes the best use of the angular resolution of Earth-based instruments, namely a fraction of an arcsec, in studying phenomena close to the nucleus. The nucleus itself is unresolved, since a 7 km nucleus at 1 AU would subtend only 0.01 arcsec. On this scale, certain comets exhibit frustratingly flat and structureless images, whilst others reveal jets originating in the central region and evolving in time (cf. Figs. 3.8 and 3.9).

Some of these jets are easily observed, even in visual observations. Others are only revealed as a result of sophisticated image processing. On the scale on which they are observed, between several hundred and several thousand kilometres, the curving effect due to solar radiation pressure

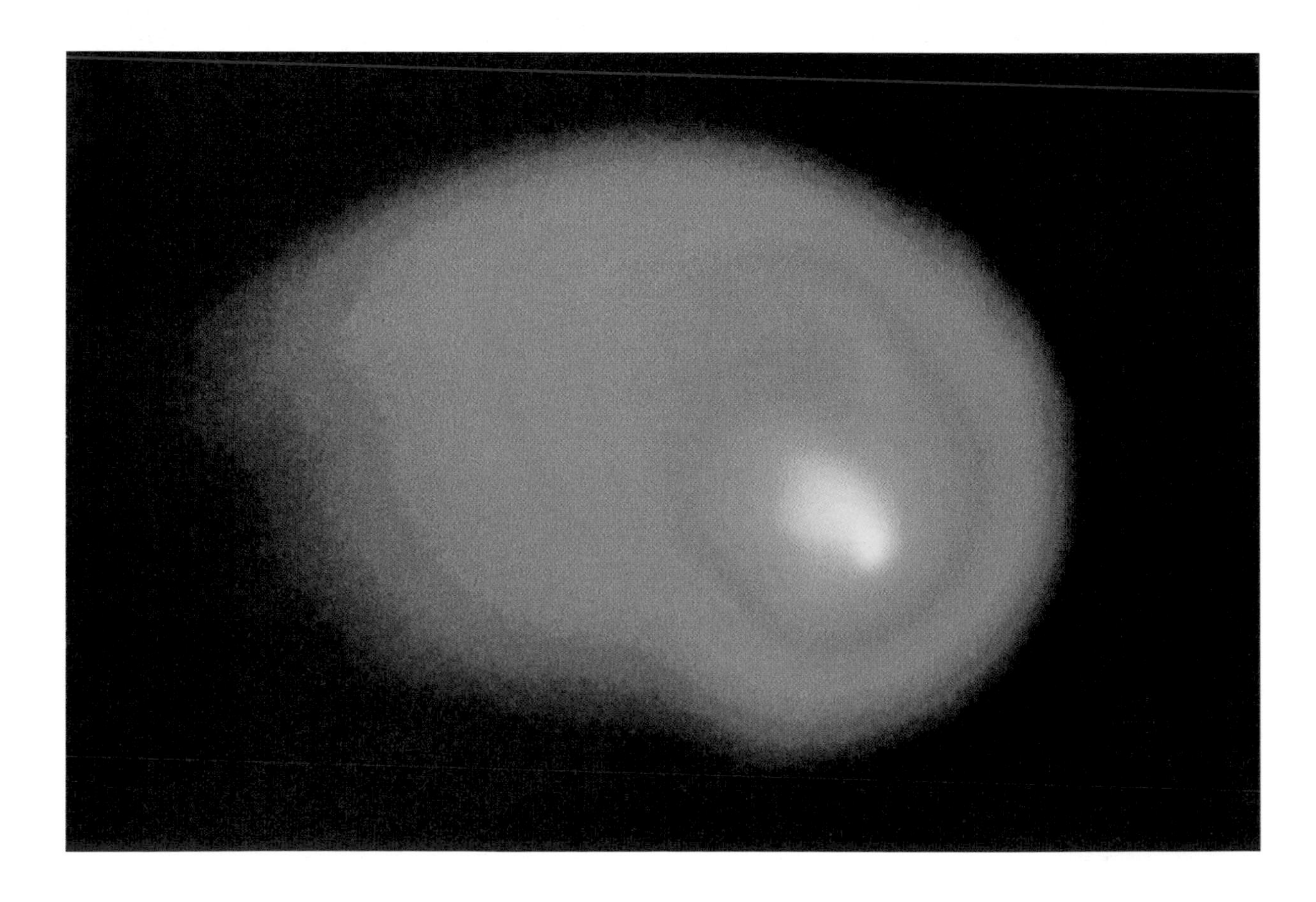

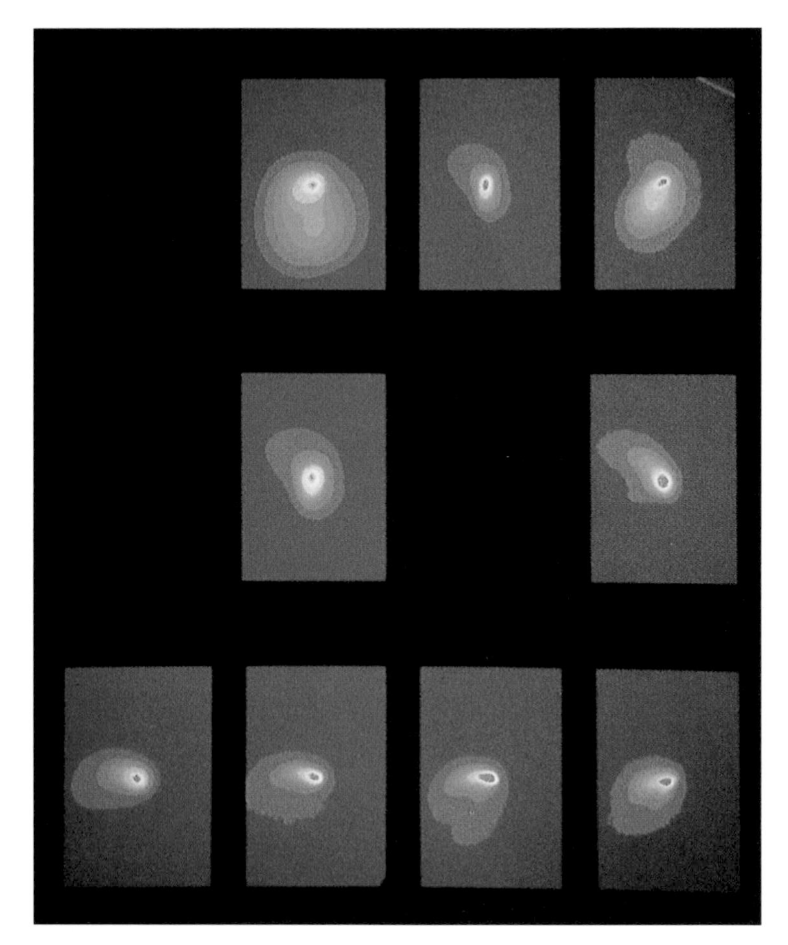

Figure 3.9.
Dust jets from comet P/Swift–Tuttle. Each of the sequences of images (of which only a few extracts, taken between 20 and 28 November 1992, are given here) shows identical jets every 2.7 days. This suggests that the cometary nucleus is spinning with this period. The same period of rotation was deduced by analysing drawings made in 1862. Images made at Pic-du-Midi observatory, courtesy of F. Colas, L. Jorda and J. Lecacheux.

Figure 3.7. opposite
True colour image of comet Halley, taken on 1 March 1986 with the ESO's 40 cm telescope. This composite image was obtained from three black and white images taken with different filters. Because of their motion between the three exposures, stars appear as trails of light of different colours. The blue plasma tail and red dust tail are revealed. The field covers 1°. Courtesy of ESO.

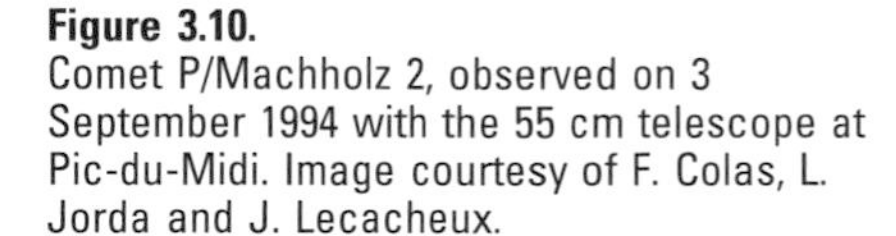

Figure 3.10.
Comet P/Machholz 2, observed on 3 September 1994 with the 55 cm telescope at Pic-du-Midi. Image courtesy of F. Colas, L. Jorda and J. Lecacheux.

Figure 3.8. opposite
Dust jets from comet P/Swift–Tuttle. Image obtained on 20 December 1992 with the 1 m telescope at Pic-du-Midi, and computer processed. Field stars have been subtracted and a digital filter applied to bring out contrast in the jets. Image courtesy of F. Colas, L. Jorda and J. Lecacheux.

Dynamics of dust tails

There can be no doubt that the tail of a comet is its most spectacular feature. It was realised quite early on that these tails develop in the antisolar direction. The German scientist Friedrich Bessel (1784–1846) was the first to suggest that they might be made up of particles emitted by the comet itself, subject to some repulsive force from solar radiation.

Dust and ion tails have quite different properties and therefore manifest themselves in quite different ways. Let us consider here the parameters and forces governing the kinematics of a particle in the dust tail.

- The particle leaves the central region with an initial speed v_0 due to the momentum of cometary gases (see Chapter 7). This speed, a fraction of a km/s, is smaller as the particle is heavier.
- The gravitational attraction of the nucleus is negligible.
- The force of attraction due to the Sun's gravity is proportional to m_g/r_h^2, where m_g is the particle mass.
- The repulsive force due to solar radiation pressure is proportional to $\pi a^2/r_h^2$, where πa^2 is the geometric cross-section of the particle.

Solar gravitation and radiation pressure exert forces in opposite directions, but it is the latter which dominates for particles of radius less than 0.6 μm (assuming a density of 1 g/cm^3), and such particles predominate in the visible cometary tail. These particles then follow hyperbolic paths relative to a frame in which the Sun is stationary.

The parameters here depend on particle size and density. It follows that cometary particles, which have a wide range of sizes and densities, will move along quite different trajectories. Hence the great width of cometary tails.

- All particles leaving the comet with the same parameters (i.e. initial velocity and solar acceleration), but at different instants, form a curve called a *syndyne*, and it is the fanning out of these syndynes, for particles of various properties, which forms the observed tail.
- All particles leaving the comet at a given instant form what is called a *synchrone*. Such a structure can sometimes be observed in the tail when the comet is subject to bursts of activity and suddenly emits a cloud of matter.
- Close to the nucleus, and viewed in a frame in which the nucleus is stationary, particles describe parabolic paths curving away from the Sun to a first approximation. Such is the fountain model devised by Eddington (see Fig. 3.11).

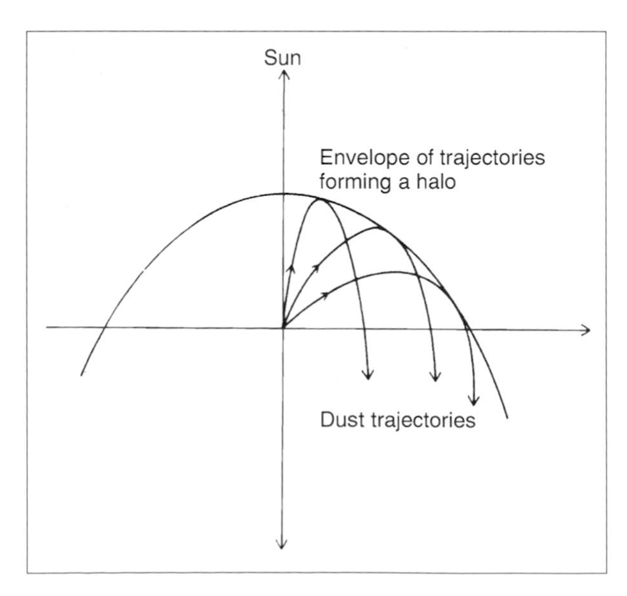

Figure 3.11.
Intermediate scale images. Dust trajectories and halo formation.

is barely apparent. On the whole these jets are directed towards the Sun, demonstrating that cometary activity tends to develop on the side of the nucleus heated by the Sun. By following changes in orientation of these jets over time, it is sometimes possible to expose a spinning of the nucleus and to deduce its period, or even the direction of its axis (see Fig. 3.9).

On a longer scale (several tens of thousands of kilometres), radiation pressure becomes much more significant and dust jets reorganise themselves to form the beginnings of a tail. According to the fountain model, proposed by Arthur Eddington (1882–1944) at the beginning of the century, the envelope of these jets is sometimes clearly apparent, forming a structure which could be interpreted as a halo (see Fig. 3.11).

It is important to distinguish structures due to gas from those due to dust, when analysing

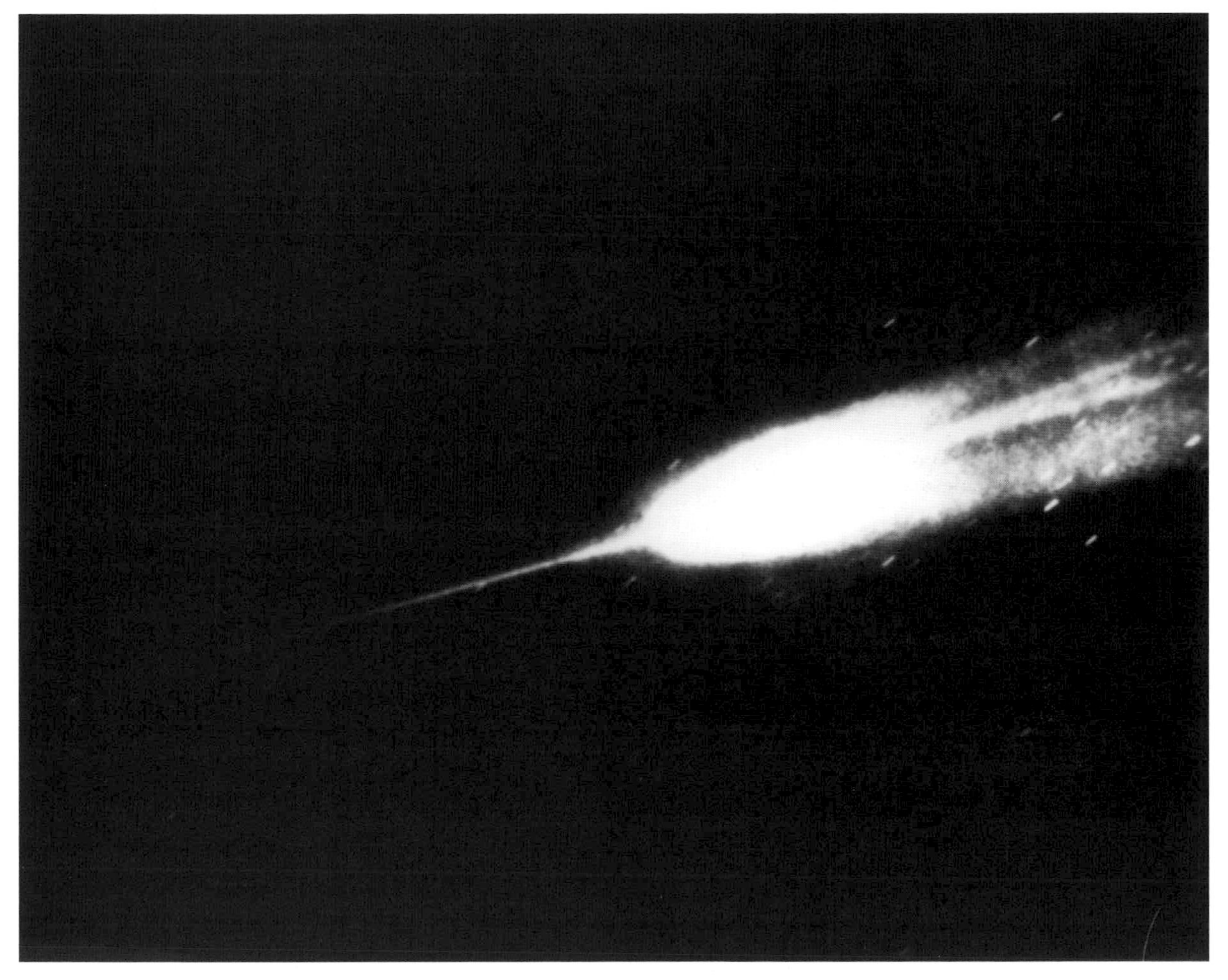

Figure 3.12.
Comet Arend–Roland 1957 III and its antitail, on 25 April 1957. In contrast with the normal appearance of the tail, the antitail points towards the Sun. This rare phenomenon occurs when the Earth lies in a comet's orbital plane. It is caused by large dust grains which left the nucleus a long time previously, and whose curved paths appear in this way through a trick of perspective. Courtesy of Paris-Meudon Observatory, photo by C. Bertaud and G. Bertaud.

cometary images. This can be done by using suitable filters, which select either *continuum emission* from dust, or emission bands of certain radicals (see Chapter 4). It was believed for a long time that only dust could exhibit jet structures. However, certain radicals, in particular CN, have revealed spatial structure in some comets, such as P/Halley. Direct observation of the spatial distribution of parent molecules will be discussed in Chapter 6.

4
Visible and ultraviolet spectroscopy

The observations described so far cannot even distinguish between gas and dust from the comet, let alone reveal the chemical composition of this material. In order to accomplish this task, a more subtle analysis of cometary light is required, known as spectroscopy.

4.1 Classical cometary spectroscopy: an historical view of visible spectroscopy

Spectroscopic research in astronomy began in the nineteenth century with visual observations using simple prism spectrometers, soon to be supported by photographic methods. Just one hundred years ago, in the Observatoire de Meudon near Paris, cometary and stellar spectra were being compared with laboratory reference spectra. This marked the beginning of *astrophysics* in the recently founded establishment. The new discipline brought together observational astronomy, experimental physics and, of course, theoretical calculations.

The most intense cometary spectral bands were observed as early as 1864 in comet Tempel 1864 II (by Giovanni Donati in Florence), in 1868 in comet Winnecke 1868 II (by William Huggins in England), and then in the Great Comets 1881 III and 1882 II (by William Huggins and Henry Draper, using photography for the first time). These were the yellow, red and blue bands of the *Swan system*, obtained experimentally in spectra of electrical discharges across hydrocarbon gases, and identified as the spectrum of the C_2 radical; and also the system of violet bands supposed at the time to be the cyanogen (C_2N_2) spectrum, but later proved to result from the CN radical. Other spectroscopic emissions were also noticed, but could not be identified. The characteristic bands of the C_3 radical, already spotted at this time, were not identified until about 1950.

These emission line spectra are superposed upon a continuous spectrum, in which can be distinguished the narrow lines of the solar spectrum. This demonstrates that solar radiation is reflected by cometary material, a result already established in 1819 by François Arago (1786–1853) [15], who observed the polarisation of cometary light.

Spectroscopic observations were subsequently improved by means of various new techniques, such as those using objective prisms around 1900. These were well suited to observing faint, extended objects. Another device was the slit spectrograph, which could clearly separate spectra from the nucleus, the extended coma and

Figure 4.1. opposite
Comet West 1976 VI. This false colour image was produced by computer processing. Image by S. Koutchmy, courtesy of CNRS, Institut d'astrophysique de Paris.

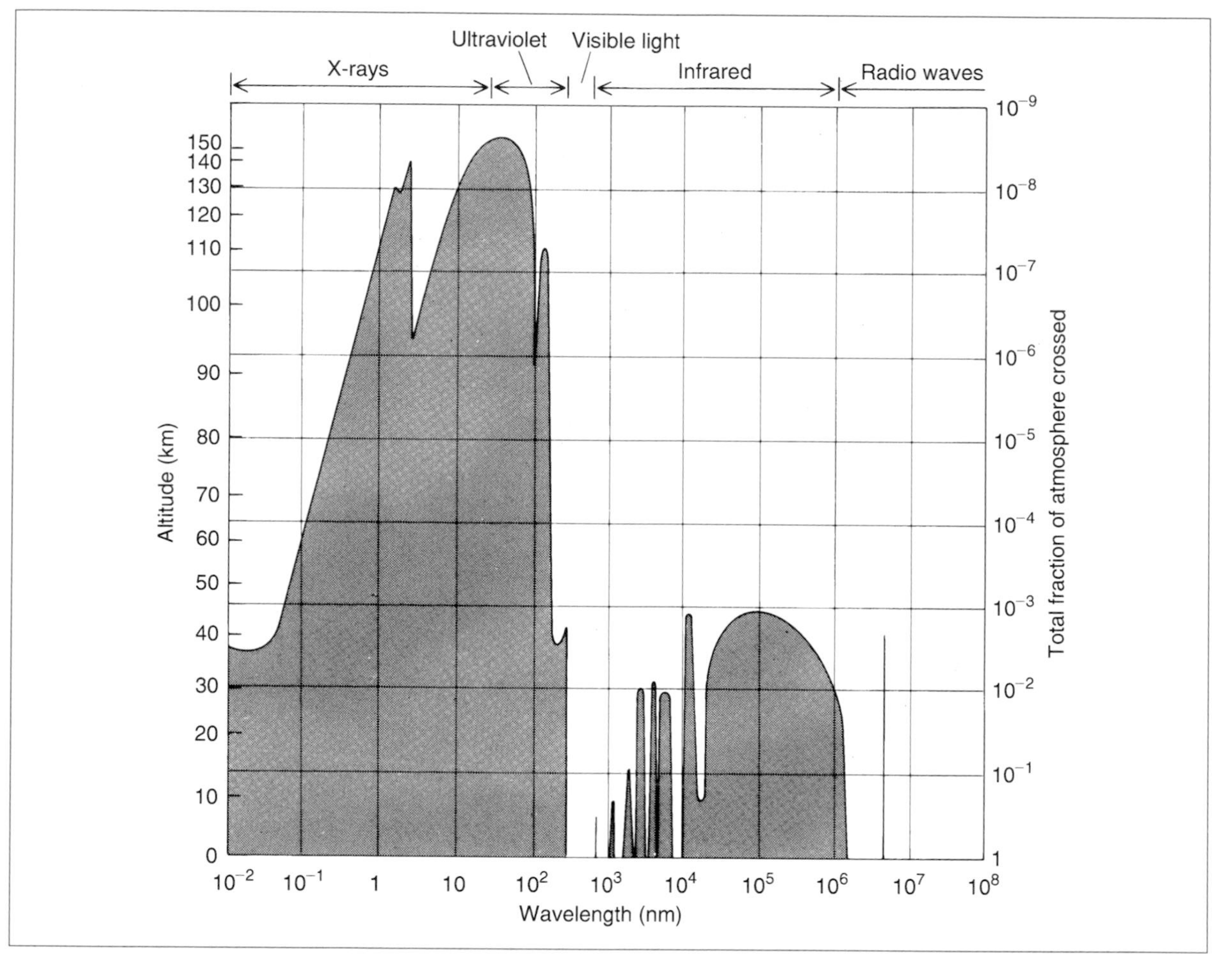

Figure 4.2.
Atmospheric windows and the various spectral regions. Upper limits of shaded areas indicate the altitude at which incident radiation intensity has been halved for the corresponding wavelength. From Bahcall *et al.* 1982 [24].

the tail, whilst making accurate wavelength measurements.

Cometary spectra exhibit mainly molecular emission bands and thereby distinguish themselves from the usual case in astrophysics. Indeed, molecular signatures are rare in stars, and although planetary atmospheres may be rich in molecular bands, they are usually from absorption, at least in the visible region.

In the visible region, cometary spectra include many bands and lines due to radicals, that is, fragments of molecules, such as CN, C_2, C_3, CH, OH, NH, NH_2, and so on. There are also lines due to atoms, such as H, O, and Na, and molecular ions, such as CO^+, CO_2^+, N_2^+ and H_2O^+. The latter molecular ion was only identified in 1974, no laboratory spectrum being available for comparison before this date. A complete list is given in Table 4.1. Identification of all these lines has represented a considerable challenge to several generations of astrophysicists and molecular spectroscopists, and many emissions remain unidentified even today.

Table 4.1. Radicals, ions and atoms observed in cometary spectra.

Radical, ion or atom	Spectral region
Radicals	
CN	Visible, near IR, radio
C_2	Visible, UV, near IR
C_3	Visible
CH	Visible, IR
OH	Near UV, IR, radio
NH	Visible
NH_2	Visible, IR
CS	UV
SO	Radio
Molecular ions	
CH^+	Visible
OH^+	Visible
H_2O^+	Visible
CO^+	Visible, UV, radio
N_2^+	Visible
CO_2^+	Visible, UV
HCO^+	Radio
H_3O^+	Radio
Atoms	
H, O	Visible, UV
C, S	UV
Na, K	Visible
Ca, Cr, Mn, Fe Ni, Cu, Co, V	Visible (only for sungrazers)
Atomic ions	
C^+	UV
O^+	UV
Ca^+	Visible

Molecular ions occur mainly in the ion tail where they are formed by interaction with ions in the solar wind. Any brightness in the tail is entirely due to radiation by its ions. The sodium D lines, forming a doublet at 589 nm, are observed in many comets within 1 AU from the Sun. They were observed to form a new kind of tail in comet Hale–Bopp (see box entitled *The sodium tail of comet Hale–Bopp*). The lines of other metallic atoms (e.g. Ca, K, Fe, Ni, Cr) only appear when the comet approaches the Sun sufficiently closely for these refractory elements to volatilise. The best example was comet Ikeya–Seki 1965 VIII, which passed the Sun at a distance of only 0.008 AU in 1965.

All these atoms and radicals, of which the most complex are C_3 and NH_2, are merely fragments. Their presence in a free state within the cometary nucleus would be difficult to explain. Hence the hypothesis that stable and volatile molecules contained in the nucleus are first sublimated, then progressively broken down by ultraviolet radiation from the Sun: *parent molecules* dissociate into the products which are observed.

Emission in most of these bands and lines is explained by fluorescence stimulated by solar radiation. Solar photons can be absorbed by atoms or molecules provided they correspond to the wavelengths which characterise those atoms or molecules. They are then re-emitted, either at the same wavelengths, or else at longer wavelengths (see Section 4.3.4). Such emissions are therefore proportional to both the solar radiation intensity and the number of cometary atoms and molecules. Certain emissions may also be caused by direct production of atoms or radicals in an excited state, through a process of photodissociation (see Section 4.4). Much work was carried out on the theoretical side by Pol Swings (1906–1983) and the Liège astrophysics group, from the 1940s. Even the subtlest details in high resolution spectra can be interpreted for certain cases, such as radicals CN, OH and C_2, using sometimes highly sophisticated models (see Fig. 4.4). Spectra depend on the heliocentric speed of the comet and thus vary with time; this is the *Swings effect* (see Section 4.3.4), which must be taken into account in theoretical models. There is excellent agreement between observed and calculated spectra for radicals CN, OH and C_2, among others. This indicates the high level of

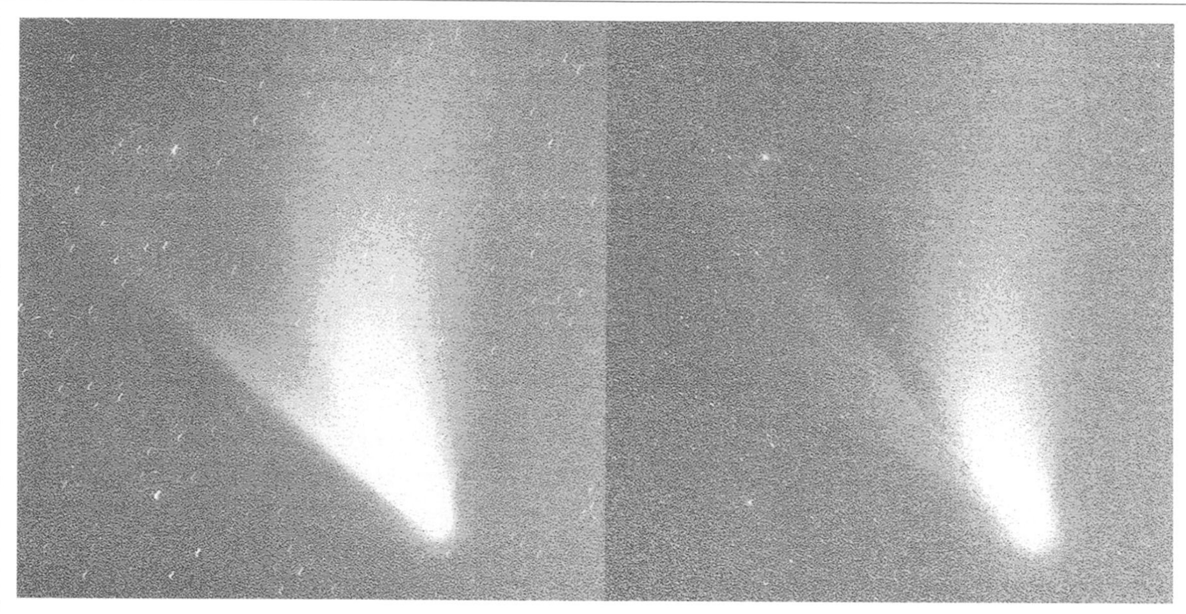

Figure 4.3.
Sodium tail of comet Hale–Bopp. The *left-hand image* was obtained with a filter centred on the sodium lines and the *right-hand image* with a filter centred on a band of the H_2O^+ ion. Both images show the broad, curving dust tail. The sodium tail is visible in the *left-hand image*, and the ion tail in the right-hand image, both as narrow and linear trails, but appearing at different angles. Observed on 16 April 1997 with the INT telescope at La Palma. From Cremonese *et al.*, 1997 [36].

The sodium tail of comet Hale–Bopp

One of the most spectacular outcomes of observations of comet Hale–Bopp was the discovery of a sodium tail. The presence of sodium in cometary comas has been known for a long time from observation of sodium yellow lines (these lines at wavelengths 5892 and 5898 Å are known as the sodium D lines, and are responsible for the emission of sodium-vapour lamps). The fluorescence rate of these lines is very high: 15 photons per second at 1 AU from the Sun (i.e. 150 times higher than the fluorescence rate of the CN radical). During the fluorescence mechanism, photons from the Sun are absorbed and subsequently re-emitted in random directions; there is thus a transfer of momentum to Na atoms, which are accelerated away from the Sun and can rapidly attain velocities greater than 100 km/s.

The sodium tail in comet Hale–Bopp was discovered from images taken with very narrow filters adapted to the wavelength of the Na D lines (Fig. 4.3). This sodium tail is observed to extend over 50 million kilometres. There is no doubt that such tails were already present in previous comets, but that they went practically unnoticed. The direction of the sodium tail is exactly opposite to the Sun, because Na atoms are accelerated by solar light. This direction differs from that of the ion tail, which is dragged along by the solar wind at a different angle, and also from that of the dust tail, subject to different dynamics. Thus, the sodium tail is really a third type of cometary tail. In addition, spectroscopic observations with very high spectral resolution have allowed observers to measure the velocity of sodium atoms as they move along the tail, thereby confirming the nature of the acceleration mechanism.

The origin of cometary sodium is still a puzzle. Only a very small fraction of the sodium within the nucleus is released in the tail, most of it probably remaining in refractories. It may be that sodium-bearing parent molecules are released from nuclear ices along with other volatile molecules, but the exact nature of these molecules is still unknown.

understanding attained with regard to the spectroscopy of these molecules and their excitation processes (see Fig. 4.4). However, the same analysis remains to be carried out for the spectra of many other molecules.

In order to obtain high resolution cometary spectra, large telescopes and sophisticated instruments must be used. This is possible only for the brightest comets, when observing conditions are sufficiently good. Wide band spectrophotometry is a less refined technique and easier to use. It is ideal for statistical study of large numbers of comets, including faint ones, and it is also well suited to following their evolution over time. The technique consists in measuring emission intensity over wide spectral bands. These are specially chosen to cover either characteristic molecular bands, like those of the radicals OH, CN, C_2 and C_3, or the molecular ions CO^+ and H_2O^+, or alternatively, spectral regions lacking in such bands, where emission from cometary dusts can be measured. Filters corresponding to these bands have recently been standardised (IHW filters, defined during the Halley observational campaign) in order to facilitate comparisons between measurements made at different observatories. The spectral ranges of these filters are shown in Fig. 4.5.

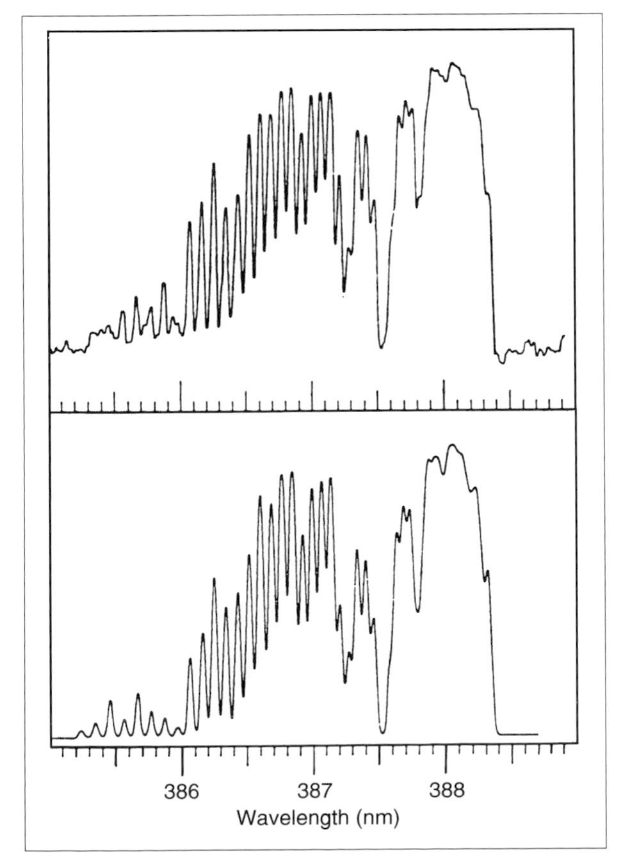

Figure 4.4.
Comparison between the spectrum of the CN radical observed around 387 nm in comet Ikeya 1963 I (*upper*) and the spectrum predicted on the basis of an excitation model (*lower*). From Malaise, 1970 [64].

Such high throughput filters allow wide field measurements to be made. This is to be contrasted with high resolution slit spectrometers which only collect a small fraction of cometary light. They can thus be used for quick and accurate measurements even when the source is faint. Molecule and dust production rates can then be assessed by standardised procedures and become the basis for statistical studies. These will be described later (Chapter 9).

4.2 Ultraviolet cometary spectroscopy

As Earth's atmosphere is opaque to ultraviolet radiation, cometary ultraviolet spectra remained hidden from us until some means of observation could be placed beyond the atmosphere. This was first attempted in 1965 for comet Ikeya–Seki 1965 VIII using a sounding rocket, and then successfully for comet Tago–Sato–Kosaka 1969 IX using the Orbiting Astronomical Observatory-2 (OAO-2), and subsequently for comet Bennett 1970 II using the Orbiting Geophysical Observatory-5 (OGO-5). Later, comets Kohoutek 1973 XII and West 1976 VI

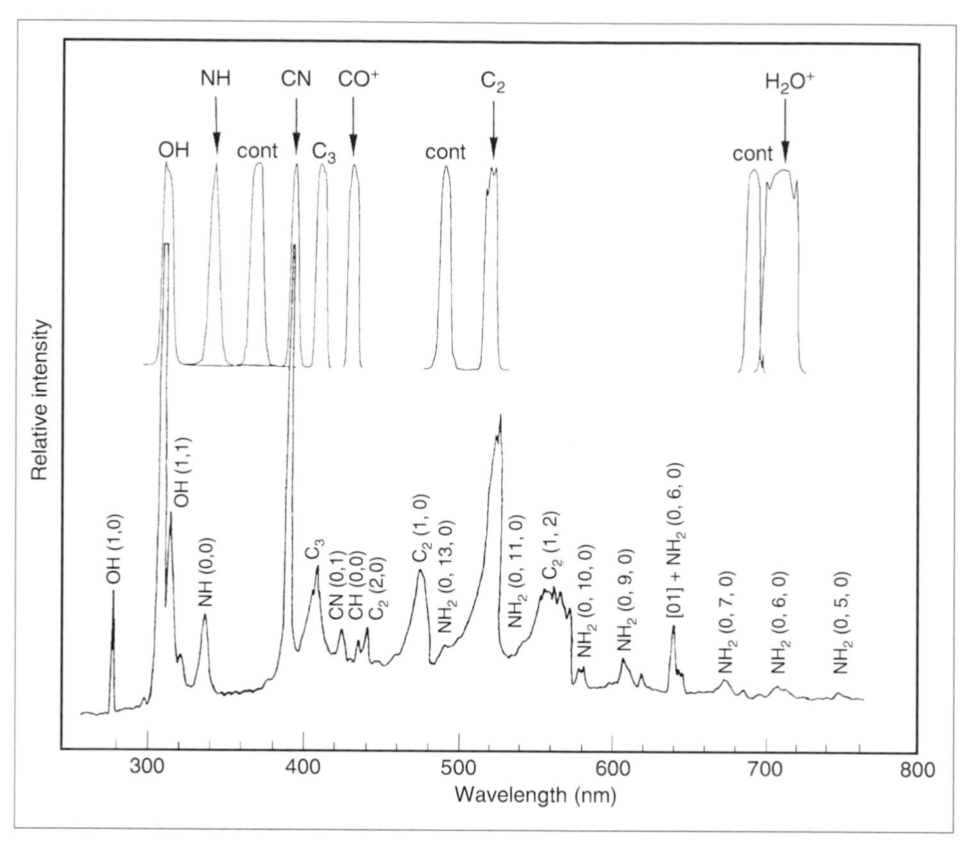

Figure 4.5.
Passbands of filters used in cometary spectrophotometry (upper), and the spectrum of a comet (lower), showing the main molecular bands. From Osborn *et al.*, 1990 [71].

were observed using rockets again. The International Ultraviolet Explorer (IUE), result of collaboration between NASA and the ESA, was launched in 1978 with a projected lifespan of three years. However, it achieved a record eighteen years of service during which it observed the UV spectra of several dozen comets. Other such observations made from the space shuttle, the Vega probes (near UV spectrum of P/Halley), and the Soviet satellite ASTRON should be mentioned. Today it is the Hubble Space Telescope (HST) which covers UV observations, although time is extremely limited owing to competition between the many users. This means that observations are not as systematic and regular as those carried out by IUE. One feature of the UV region is that the Sun is not a great producer of UV radiation, so that the continuum flux reflected off cometary dust is very weak, and in fact, almost undetectable. Fluorescence rates are weak for the same reason (see Section 4.3.4), despite the fact that molecular electronic bands and atomic lines are often intense in the UV region. The UV spectrum of comets is dominated by the Lyman α line of the hydrogen atom at 121.5 nm, and also the OH radical electronic band near 300 nm.

The Lyman α line corresponds to a fundamental emission of the hydrogen atom and is also an intense emission line from the Sun. The

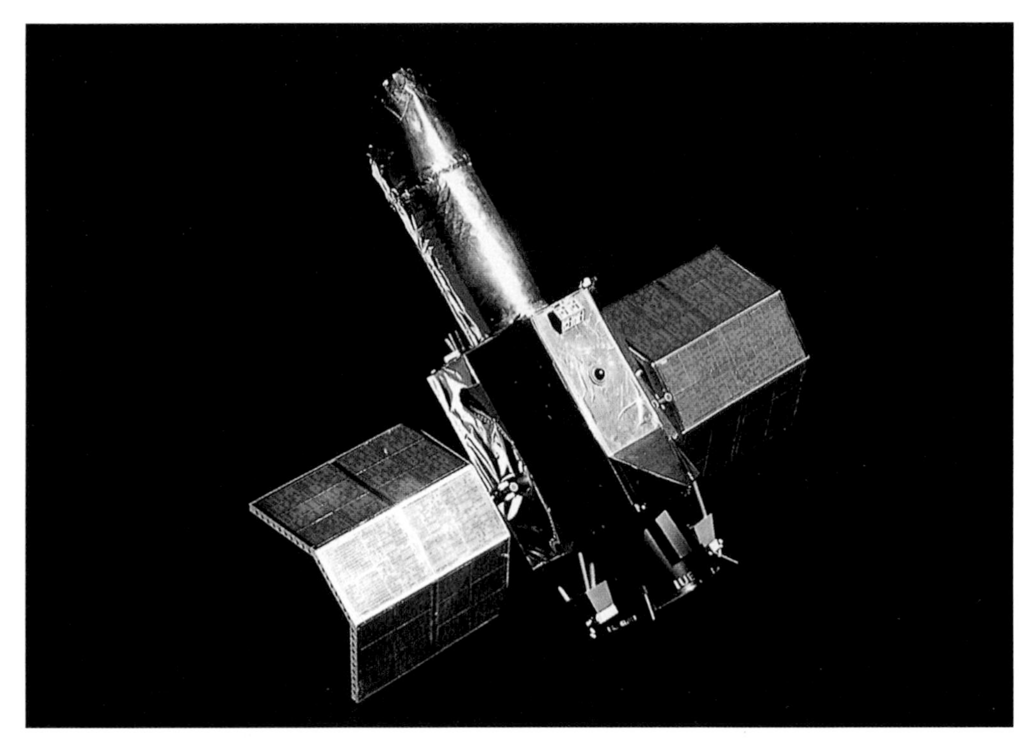

Figure 4.6.
The IUE satellite, launched in 1978 and in operation until 1996, observed the UV radiation of many sources, by bypassing the barrier of the Earth's atmosphere. It thus measured the UV spectra of more than forty comets. Courtesy of the ESA.

latter is an efficient cause of excitation by the process of resonant fluorescence. The first observations of this line in relation to comets were to reveal an enormous cloud of hydrogen, extending over about ten million kilometres. The great size of this *hydrogen envelope* is due to the long lifetime of the hydrogen atom. Its lifetime is limited only by charge exchange with ions in the solar wind, which can ionise it, and its high speed (around 20 km/s), resulting from the large excess of energy produced in photodissociation of H_2O and OH (see Section 4.4).

The most abundant cometary radical is undoubtedly OH, for it originates mainly from H_2O. This radical possesses a system of electronic bands around 300 nm which, like its system of radio lines at 18 cm, gives an indicator of the comet's gas production rate and provides a way of following the level of activity. The 300 nm bands lie in the near UV, just at the limit of what can be observed from Earth.

A further important signal in the UV region is that of the *fourth positive system* of carbon monoxide, CO, at around 145 nm. Fluorescence from this system, also observed in the upper atmosphere of Venus, is a well understood process. In contrast to other atoms or molecules observed in the visible or in the UV, CO is a stable molecule which can originate in cometary ices. This molecule was detected for the first time in comet West 1976 VI, from sounding rocket observations. An abundance of 20% relative to water was discovered, which is remarkably high. (As water is considered to be the main volatile component in cometary nuclei, the abundance of other cometary components is usually given relative to that of water. In the present work, these abundances will always be given as a relative number of molecules, rather than as a relative mass, with respect to water.) Subsequently, the IUE satellite effected many UV observations of comets in these CO bands, but revealed abundances of only a small percentage at best, even in the brightest comets. This inconsistency was only later understood, following the Halley campaign, during which CO was observed simultaneously by sounding rocket, by IUE, and by mass spectrometers on board space probes (see Chapter 5). It seems that CO is produced only in part by the cometary nucleus, and in greater proportions by some extended source in the coma. IUE, which has a field of view of only a few seconds of arc, was observing mainly the CO produced in the nucleus, whereas rocket-based observations had a much greater field of view and thus took in some of the extended source, thereby detecting far higher levels.

A result of fundamental importance obtained with the Hubble Space Telescope has been identification of the so-called *Cameron bands* of CO, located around 220 nm, in comet P/Hartley 2 (see Fig. 4.7). These bands have an extremely low rate of fluorescence and consequently are said to be *forbidden bands*. They would appear to be caused by production of

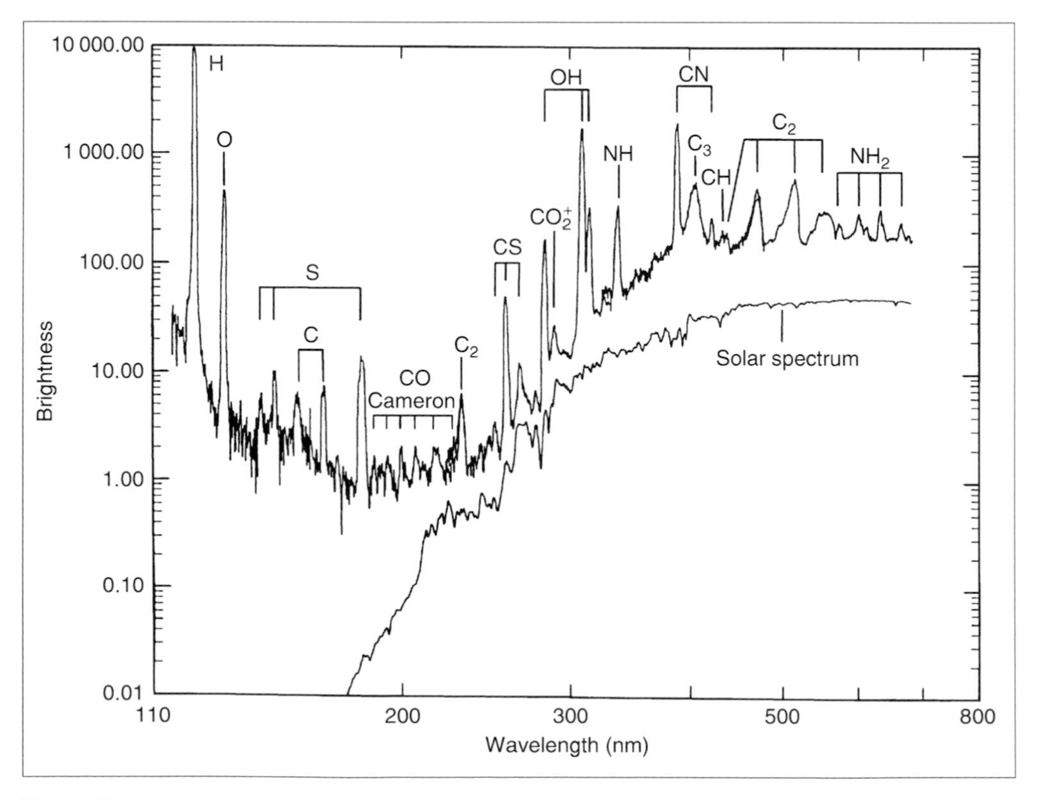

Figure 4.7.
The spectrum of comet P/Hartley 2, observed by the Hubble Space Telescope in the visible and ultraviolet regions. From Weaver and Feldman, 1992 [79].

electronically excited CO resulting from photodissociation of CO_2 (see Section 4.4). Here then was a means of observing CO_2 and determining its rate of production, independently of measurements relating to CO_2^+ bands, which do not lend themselves to quantitative analysis. The approach also avoided reference to CO_2 infrared bands, which are not accessible from Earth because there is carbon dioxide gas in the atmosphere.

Other cometary UV lines are those of the C and S atoms, which are produced as a final stage in the photodissociation of carbon and sulphur compounds. Their quantitative analysis, if it were well understood, could lead to an evaluation of the abundance of these elements in cometary ices. The radical CS has been clearly identified by its bands around 260 nm. Its most likely parent molecule is CS_2, which has a very short lifetime (about 500 seconds) and would thus be difficult to observe directly.

The molecule S_2 remains one of the great mysteries of cometary chemistry. It was identified without any doubt in comet IRAS–Araki–Alcock 1983 VII when its near UV lines were detected by the IUE, and recently in comet C/1996 B2 (Hyakutake). However, it has never been observed in any other comet. Comets IRAS–Araki–Alcock and Hyakutake both passed exceptionally close to Earth and this would certainly have facilitated detection of S_2, a molecule with a very short lifetime. It is not known what might be the origin of this molecule. Sulphur will only sublime at relatively high temperatures, which would not generally be attained on cometary surfaces or cometary grains. Moreover, the equilibrium form of its vapour is S_8 rather than S_2. The origin of sulphur in these comets thus remains enigmatic.

4.3 Molecular spectroscopy and the spectra of comets

Molecules and atoms absorb and emit light at well-defined wavelengths, those corresponding to their spectral lines. The theory of quantum mechanics gives an extraordinarily precise account of this phenomenon. Each molecule is associated with its own characteristic spectrum, so spectroscopy has become an ideal tool for non-destructive chemical analysis, as well as for analysis at a distance. The advantages for astrophysicists are obvious.

According to quantum mechanics, the energy states of a molecule are not distributed over a continuum, but rather are quantised, taking on a discrete set of values. When the molecule goes from one state to another, a photon of well defined wavelength is either emitted or absorbed. Several types of energy must be taken into account for a molecule, due to its rotation, the vibrations of its constituent atoms, or the motion of its electrons.

4.3.1 Rotational energy and radio lines

A molecule has rotational energy when it is rotating about its own axis, rather like a spinning top. Pure rotational lines, which correspond to a change of rotational state, lie in the radio region of the spectrum. Lifetimes of these rotationally excited states are typically of the order of a few seconds, but decrease very rapidly as the energy level gets higher.

In the simple case of a diatomic molecule such as CO, the rotational energy levels can be numbered by a quantum number J and the energies of the various levels are given by

$$0,\ 2B,\ 6B,\ 12B,\ \ldots\ ,\ J(J+1)B,\ \ldots\ ,$$

where B is a constant proportional to the moment of inertia of the molecule. Only transitions between consecutive energy levels are allowed; thus $J \longrightarrow J+1$ occurs through absorption of a photon and $J+1 \longrightarrow J$ occurs through emission of a photon. The energy changes, from which line frequencies can be calculated, therefore take values from

$$2B,\ 4B,\ 6B,\ \ldots\ ,\ 2JB,\ \ldots .$$

In the case of the CO molecule, rotational lines are indeed observed in the millimetre and submillimetre wavelengths of the radio region, at frequencies of 115 GHz, 230 GHz, 345 GHz, and so on. For polyatomic molecules, such as H_2O, the situation is much more involved; two or more quantum numbers are now required in order to number energy states, and they no longer follow a simple one-dimensional sequence.

It should also be mentioned that only those molecules possessing an electric dipole moment can emit rotational lines. The intensity of these lines is proportional to the square of this dipole moment. Symmetric molecules, such as CH_4 or CO_2, have zero electric dipole moment and thus produce no rotational lines. They cannot therefore be observed in the radio region.

4.3.2 Vibrational energy and infrared lines

A molecule has vibrational energy when its constituent atoms are vibrating relative to each other. A change of vibrational state gives rise to infrared lines. Lifetimes of vibrationally excited states are of the order of a hundredth of a second. Many vibrational modes may exist, depending on the complexity of the molecule. For example, there are stretching modes which correspond to a variation of the distance between atoms. The stretching mode of the C—H group corresponds to a wavelength of about 3.3 μm, and that of C—O to about 4.5 μm. There are also bending modes, related to variations in angle between bonds linking three atoms. The bending mode of H—O—H gives a line at 6.3 μm, and that of O—C—O gives a line at 15.0 μm. Since a change in vibrational energy is usually accompanied by a change in rotational energy, a *band* comprising several vibrational-rotational lines at neighbouring wavelengths is generally observed. Relative intensities of lines in such a band depend on the distribution of the various rotational levels for the molecule in question.

Each vibrational mode possesses a set of possible energy levels, numbered as

$$0,\ 1,\ 2,\ \ldots,\ v,\ \ldots .$$

The pure vibrational transitions are the fundamental bands corresponding to $v \longrightarrow v-1$, and their harmonics for which $\Delta v = 2, 3, \ldots$. For a diatomic molecule such as CO, there is only a stretching mode, whose fundamental wavelength lies at 4.5 μm.

4.3.3 Electronic energy and lines in the visible and UV

The electronic energy of a molecule refers to energy associated with the arrangement of electrons around its various atomic nuclei. The related transitions produce lines in the UV, the visible, and sometimes the near infrared. Lifetimes of excited electronic states are of the order of the microsecond or the nanosecond. Electronic transitions in molecules are accompanied by changes of vibrational and rotational energy, thus giving rise this time to a *system of bands*. Each band corresponds to a change in vibrational quantum number $v' \longrightarrow v''$ and is itself composed of a set of lines due to changes in rotational energy. Rotational structure can only be discerned in spectra measured with high spectral resolution. Needless to say, atoms and atomic ions manifest only electronic transitions.

4.3.4 Cometary spectra and fluorescence mechanisms

Cometary spectra observed in the UV, visible and infrared regions are essentially produced via the mechanism of fluorescence, driven by solar radiation. A molecule or atom first absorbs a solar photon of appropriate wavelength, taking it to some higher vibrational and/or electronic energy level. The molecule or atom then falls to a lower energy level in one or more steps, emitting one or several photons, respectively.

If the re-emitted photon has the same frequency as the one absorbed, the process is referred to as *resonant fluorescence*. This is the situation for the Lyman α line of atomic hydrogen, at 121.5 nm. However, a molecule which has been excited electronically or vibrationally by a solar photon can de-excite through various channels corresponding to different changes of rotational or vibrational energy state. A rich and characteristic fluorescence spectrum is thereby generated. An electronic excitation may also lead to a genuine cascade of fluorescent emissions, passing through a number of intermediate electronic states and hence emitting several photons at longer wavelengths than that of the exciting photon.

The rate of fluorescence g is defined as the number of photons emitted per second and per molecule, within a given band. g depends on the solar radiation available to excite fluorescence, which is proportional to r_h^{-2}, the intrinsic *strength* of the band, and the *branching ratios* between the various possible de-excitation channels. As an illustration, at 1 AU from the Sun, the rate of fluorescence of the hydrogen Lyman α line is 1.4×10^{-3} photons per second; for the violet system of CN at 388 nm, the rate is 9×10^{-2} photons per second. The fourth positive system of CO, in the UV, has a rate of only 1.2×10^{-6} photons per second.

Solar radiation responsible for exciting cometary molecules into fluorescence differs from a blackbody spectrum by the presence of strong absorption lines, known as *Fraunhofer lines*. Any interpretation of molecular band structures must take into account these absorption lines. Their wavelengths are shifted by the Doppler effect, since the heliocentric speed of the comet is often as great as several tens of kilometres per second; this is the *Swings effect*.

Apart from CO, cometary parent molecules such as H_2O and CO_2 do not have electronic bands of significant fluorescence rate g in the visible or UV. They do possess strong electronic bands, but excitation of these bands leads to destruction of the molecule before it can de-excite by photon emission. However, fluorescence can occur through excitation of the fundamental vibrational modes of these molecules, giving rates like 2.7×10^{-4} photons per second for the stretching mode of H_2O at 2.7 μm, and 2.6×10^{-3} photons per second for the mode of CO_2 at 4.2 μm.

Emission of pure rotational lines depends on the distribution of molecules over the various rotational levels. In the region of the inner coma, where densities are rather high, collisions will establish a distribution corresponding to *thermal equilibrium*, entirely determined by the gas temperature (see Chapter 7). Further from the nucleus, collisions become rarer or non-existent, and gas molecules are subject to two opposing phenomena with regard to their distribution over the various rotational levels. Firstly, there is fluorescence following excitation of electronic or vibrational states, a phenomenon which tends to populate higher and higher rotational levels. Then there is spontaneous emission of pure rotational lines, a phenomenon which tends to bring molecules back down to lower rotational levels. Having suffered a certain number of such fluorescence cycles, molecules should reach a state of equilibrium, called *fluorescence equilibrium*. In fact there is a gradual transition from thermal to fluorescence equilibrium over the region between inner and outer coma.

4.4 Photolytic processes and the spatial distribution of cometary molecules in the coma

4.4.1 Molecular lifetimes

When a molecule absorbs a photon of energy greater than its bonding energy, it may dissociate. This process is called *photodissociation*. For many simple molecules, the first excited electronic states satisfy this condition and corresponding electronic transitions are dissociative. The main wavelengths causing dissociation are to be found towards the far end of the ultraviolet, and occasionally in the near ultraviolet or the blue.

Most ultraviolet radiation is filtered out by the Earth's atmosphere before it can reach the surface of our planet. If this were not the case, the majority of organic molecules and a high proportion of inorganic molecules would rapidly photodissociate. The same is not true in cometary atmospheres, where molecules are directly subjected to solar radiation. Some resist better than others. At 1 AU from the Sun, H_2O, HCN and methanol have lifetimes of the order of 80 000 seconds, whereas ammonia and formaldehyde will last on average only 7000 and 5000 seconds, respectively. CO and CO_2 have much longer lifetimes, estimated at 1.3×10^6 seconds and 500 000 seconds, respectively.

When a molecule photodissociates, it produces other molecules, radicals or atoms. If the energy of the absorbed photon is high enough, even ions may be produced; this is known as *photoionisation*. For example, under photolysis by solar radiation, water may undergo any of the following reactions, in the given proportions:

$$H_2O + h\nu \longrightarrow H + OH \quad (85.5\%)$$
$$H_2O + h\nu \longrightarrow H_2 + O \quad (5.0\%)$$
$$H_2O + h\nu \longrightarrow 2H + O \quad (6.3\%)$$
$$H_2O + h\nu \longrightarrow H_2O^+ \quad (2.7\%)$$
$$H_2O + h\nu \longrightarrow H + OH^+ \quad (0.5\%)$$

where $h\nu$ denotes the incident solar photon.

4.4.2 Excess energy of fragments

When a photon is absorbed by and dissociates a molecule, its energy will generally be greater than that required merely to break molecular bonds, and often by a wide margin. Any excess energy will be carried off by the fragments in the form of electronic, vibrational, rotational or translational energy.

Electronic energy Some fragments may be created in an excited electronic state. These molecules or atoms will emit light when they spontaneously de-excite, in a process called chemiluminescence, which is fundamental in the emission of cometary spectra and may at times surpass fluorescence processes. A good example is provided by oxygen atoms produced when water photodissociates. These emit lines at 557.7, 630.0 and 636.4 nm, which are called *forbidden lines* because the probability of their being excited by fluorescence processes is so low. A further example was discovered only recently during observation of the spectra of P/Hartley 2 by the HST (see Fig. 4.7). Emission of electronic bands in the Cameron system of CO, around 200 nm, comes from production of excited CO following photodissociation of CO_2.

Spatial distribution of cometary molecules

The phenomena described in the text determine density distributions of parent molecules and radicals in cometary atmospheres. For a parent molecule, the density $n(r)$ at distance r from the nucleus is given, for uniform emission, by

$$n(r) = \frac{Q}{4\pi r^2 v_{exp}} \exp(-r/\gamma) ,$$

where Q is the rate at which the molecule is produced, and v_{exp} the speed of expansion of the atmosphere.

The exponential term models destruction of the molecule, $\gamma = v_{exp} t$ being the *scale length* of the molecule, and t its lifetime. For a daughter molecule, assuming it follows the same path as its parent, which amounts to neglecting its ejection speed,

$$n(r) = \frac{Q}{4\pi r^2 v_{exp}} \frac{\gamma_d}{\gamma_p - \gamma_d} \times \left[\exp(-r/\gamma_p) - \exp(-r/\gamma_d)\right] ,$$

where γ_p and γ_d are the scale lengths of parent and daughter, respectively.

These simple formulas, known as the *Haser laws*, were long used as a first approximation in analysis of cometary spectra, and to deduce molecular production rates. A deeper analysis must also include the fact that v_{exp} is not constant, but increases with distance from the nucleus (see Chapter 7); and even more importantly, the fact that daughter molecules are ejected with some additional velocity relative to their parent molecules. Moreover, radicals like CN, which are subject to high rates of fluorescence, are accelerated in the antisolar direction; this process is similar to radiation pressure on dust and tends to create a fountain distribution for these molecules, rendering the coma non-symmetric.

Translational energy This is related to the speed of ejection of the fragment. In the case of water photolysis, which has been particularly well studied, the OH radical is ejected with a mean speed of 1.1 km/s, and the hydrogen atom at 20 km/s on average. These speeds are large compared with the expansion speed of the coma as a whole, which is about 1 km/s, and the thermal agitation speeds of its molecules, which are generally a fraction of 1 km/s. The fast-moving fragments of photodissociation thus enter into collisions with other particles in the cometary atmosphere and cause *photolytic heating* (see Chapter 7).

Any more detailed study of processes involved in destroying parent molecules must take into account collisions with high energy ions in the solar wind. These are as effective as photodissociation processes in destroying molecules like CO and CO_2. Variations in solar activity cannot be neglected either; solar UV emission varies significantly over the eleven year solar cycle, as well as manifesting unpredictable short term changes. Such variations may reduce lifetimes of certain molecules by half relative to values cited above, which refer to calm solar conditions.

5
The 1986 exploration of comet Halley

5.1 Preparing the campaign of observation

We have seen that the astronomical observation of comets is faced with a fundamental paradox: either the comet is far away and hence too small and cold to be easily observed from Earth; or it is nearby and hence too active for observation of the nucleus, which is much less bright than the coma and tails. For this reason, astronomers have longed for an opportunity to observe a comet at closer quarters.

It was only in the second half of the twentieth century, with the technological advances made in space exploration, that this dream became a real possibility. In the 1970s, when the astronomical community was still revelling in the success of the Apollo missions to the Moon, several spacecraft were sent to Mars, Venus and the Jovian planets. An in-depth study of Mars was under preparation with the Viking mission. Across the Atlantic, the more ambitious of investigators began to imagine a possible encounter with a comet in space.

It remained to be decided which object should be studied. Although planetary orbits are known to great enough accuracy to allow encounters with spacecraft, this is not the case for the majority of comets. Unexpected comets can be eliminated at the outset. They give too short notice of their arrival for the launching of a space probe to be organised. The only possible targets are periodic comets, which have been observed on many previous occasions, and whose non-gravitational forces are sufficiently well known. At the end of the 1970s, there was general agreement among cometary scientists that comet Halley was the best choice. This comet had the advantage of being particularly active, as a result of its long period. It also had a perfectly known trajectory, the first of its apparitions going back to ancient times. Precise knowledge of the orbit is an essential factor when planning a space exploration. Most other well-known periodic comets exhibit much shorter periods than comet Halley, and having experienced numerous close encounters with the Sun, a large fraction of their ice has already sublimed. This means that their activity is significantly reduced. With its 76 year period, comet Halley therefore constitutes an exception. Its last apparition in 1910 had brought it very close to Earth and excited much emotion; but its 1986 apparition presented an ideal opportunity for cometary scientists.

However, from a purely geometric point of view, the 1986 apparition was particularly unfavourable (see Fig. 5.3). At perihelion passage on 9 February 1986, the comet was situated almost behind the Sun and was therefore very badly placed for Earth-based observation. The best telescopic observations were thus carried out at

Figure 5.1.
P/Halley in 1910. A remarkable photo taken at the Mount Wilson Observatory on 8 May 1910. Coma and dust tail, divided into two parts, are visible. The Sun is on the left relative to the photo. Courtesy of Mount Wilson Observatory.

the instant when the comet's geocentric distance was minimal, before perihelion in November 1985 and after perihelion in April 1986.

Concerning *in situ* observation, another constraint was imposed. As for most comets, the orbit of comet Halley does not lie in the ecliptic plane. However, considerable energy is required to send any space vehicle out of this plane. The simplest solution consisted therefore in intercepting the comet just as it crossed the ecliptic. The date chosen was the week of 13 March 1986, after the comet's perihelion passage. A group of five space probes, one European, two Japanese and two from the USSR, observed comet Halley between 6 and 15 March 1986 (see Table A.3 in Appendix). This joint programme constitutes, even today, a unique event in the annals of space exploration, laying a foundation for effective international collaboration in investigation of the Solar System.

At the beginning of the 1980s, there was fierce competition between the larger space agencies in this area, dominated by NASA after its successful Viking missions to Mars and Voyager missions to the Jovian planets. In view of the undoubted supremacy of the United States, the Soviet Union had been concentrating

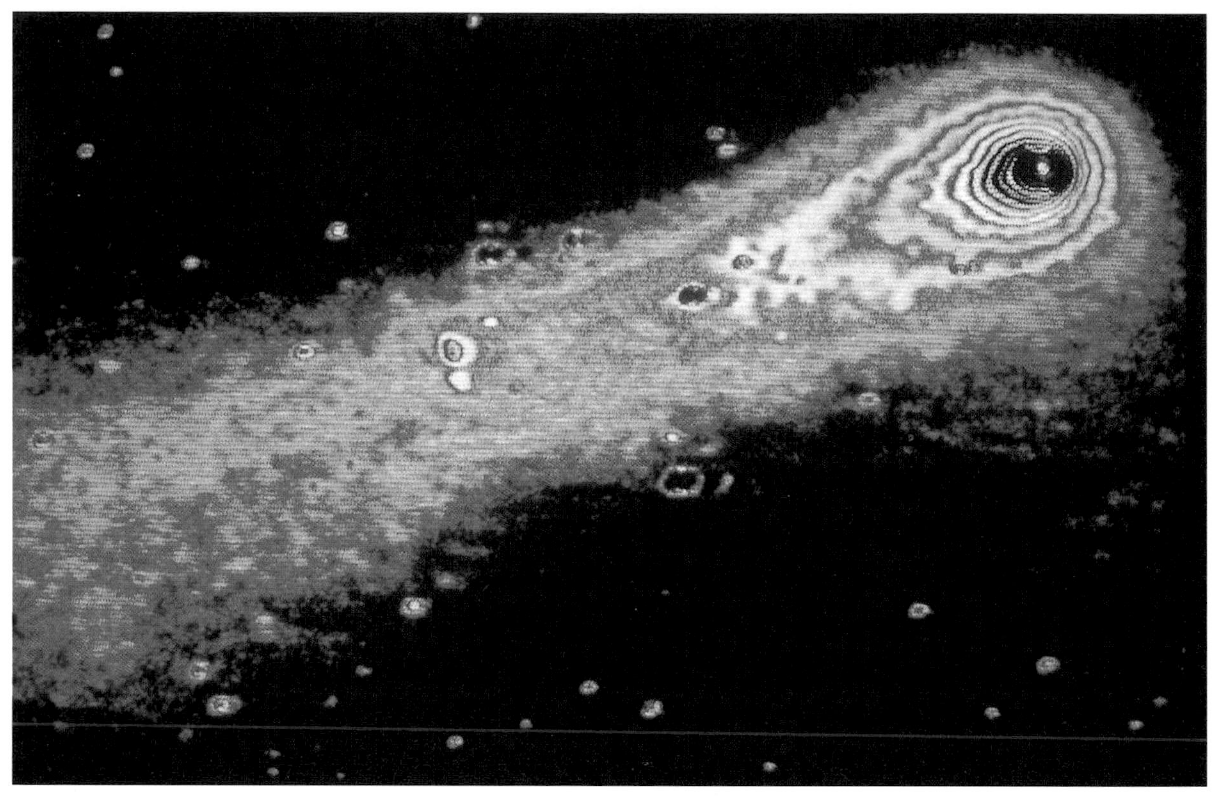

Figure 5.2.
Photo of P/Halley taken at Pic-du-Midi on 29 May 1910 and reprocessed by computer. With such processing, flux emitted at various points of the coma can be measured to greatest possible accuracy. Photo courtesy of Bordeaux Observatory, processed by the CNRS at Paris Observatory.

on the planet Venus. For its part, the new European Space Agency (ESA) had not yet launched a probe beyond a terrestrial orbit.

In such a context, NASA might have been expected to take the initiative over the conquest of a comet. Indeed, this was the hope of the American scientific community, which had been mobilised for several years. But they were to be disappointed, NASA's projects being cancelled for political and budgetary reasons. The other space agencies were thus blessed with an unexpected opportunity, and took full advantage of it. The Soviet Union decided to reuse two spacecraft designed for exploration of Venus, redirecting them towards comet Halley after releasing a module for descent into the Venusian atmosphere. In record time, Europe had defined a highly ambitious mission which aimed at a close encounter with the cometary nucleus. It was based on technology developed originally for terrestrial exploration satellites. The Japanese set up two missions whose aim was to study, at greater distances from the comet, its interaction with the solar wind. It was thus that in 1986 five space probes were simultaneously engaged in exploration of a Solar System object which lay outside the Earth–Moon system.

Comet Halley is not, however, the only comet to have been observed *in situ*. In order to make up for its failure with regard to comet Halley,

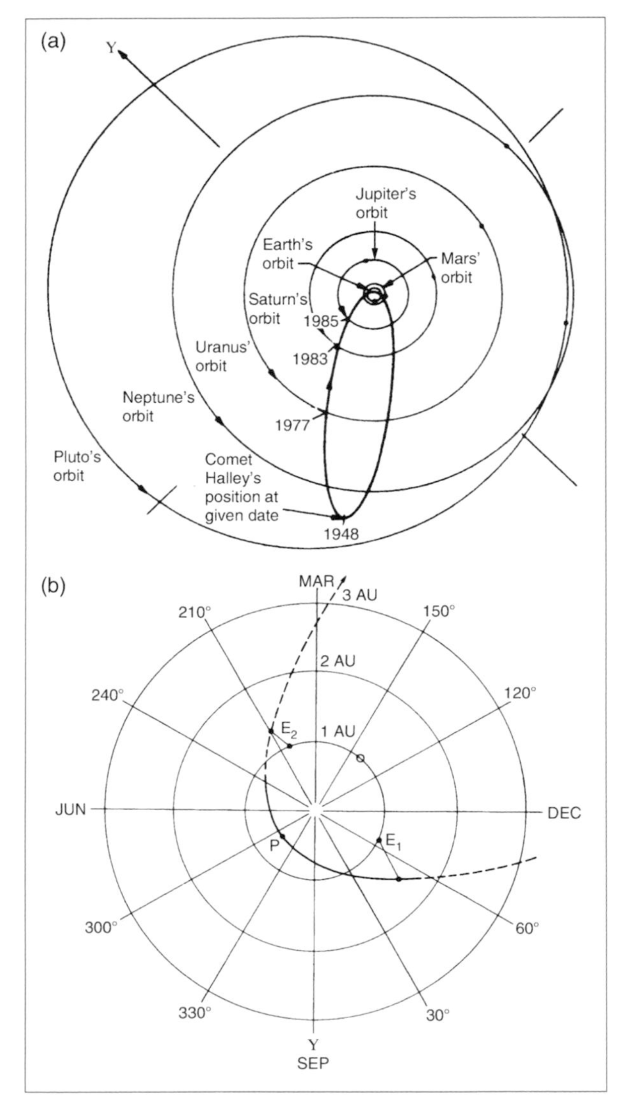

Figure 5.3.
Geometric configuration of the 1986 apparition. Trajectory of comet Halley (a) projected onto the ecliptic plane, and (b) at the time of the 1986 apparition. It can be seen that, when the comet was at perihelion (P), on 9 February 1986, the Earth was on the other side of the Sun. Points E_1 and E_2 correspond to optimal observation conditions, when the comet's geocentric distance was minimal. IHW.

NASA decided shortly before the 1986 apparition to redirect a probe which had been placed in terrestrial orbit for study of the interplanetary medium. This was the ISEE-3 mission, which was renamed the ICE (International Cometary Explorer) for the purposes of its new role. In November 1985, it passed close by the periodic comet P/Giacobini–Zinner and made a study of its interaction with the solar wind. Furthermore, following the success of the European mission Giotto to comet Halley, the ESA decided to reuse the same probe, several years after, in the study of another periodic comet, the much less active P/Grigg–Skjellerup.

5.2 The Earth-based observation campaign

Even for the 1910 apparition of comet Halley, astronomers had organised a campaign to study the comet's evolution with the greatest possible accuracy. In those days, observation was limited mainly to photographic plates. These could be used to study evolution of jets from the nucleus, as well as development of dust and ionised gas tails. It also provided some low resolution spectra. Because the 1910 apparition was particularly favourable, the comet passing at only 0.60 AU from Earth, an impressive collection of photos has been preserved. These can be used to retrace, sometimes from one day to the next, the activity of the comet, both before and after its perihelion passage on 20 April 1910. The comet was observed from September 1909 to June 1911. The geometric configuration was so propitious, and the quality of the photographic plates so good, that this data has recently been reanalysed by computer technology. The plates were digitised in order to be given a more quantitative interpretation. This led to a better description of the temporal evolution of the inner coma and dust jets, close to the comet's perihelion passage of April 1910.

Considerable progress has been made in astronomical instrumentation since the beginning of the century. Whereas spectroscopy was then limited to the visible region and could only identify products of photodissociation and photoionisation, this technique has recently been extended to the other wavelength regions, i.e. ultraviolet, infrared and radio, hence allowing detection of other molecules, in particular the elusive parent molecules originating directly from the cometary nucleus. Instrumentation related to imaging has been developed in parallel, notably with the development of CCD-type digitised cameras.

Equipped with these new means, and engaged in a quite unprecedented programme of space exploration, the motivation of the astronomical community to organise an international campaign of telescopic observation could not have been greater. It was NASA, excluded against its will from the space programme, which took up the challenge of coordinating this campaign, under the name of International Halley Watch (IHW). Covering a wide range of sub-disciplines (astrometry, nuclear imaging, wide field imaging, UV and visible spectrometry, infrared observation, radio observation, coordination with amateur astronomers, and so on), the task of this organisation was to instigate and encourage observation programmes in the various astronomical centres, to transmit information and coordinate efforts, and ultimately to collate the whole set of data in archives. Led by Ray Newburn and Jurgen Rahe (1940–1997), IHW was a remarkable success, and the archives (two dozen CDROMs) were distributed to the scientific community in 1993. Several years before the passage of comet Halley, a somewhat less bright periodic comet, comet Crommelin, had been chosen for a general rehearsal. This was a positive experience, as much for coordination of observations as for archiving; but it ought to be said that, for this much fainter comet, far fewer astronomers had been mobilised, and the volume of data collected was considerably smaller.

In the 1970s, new and more efficient detectors were beginning to compete with the photographic plate. At the same time, a new type of camera was being developed, using CCDs, already mentioned in Chapter 3.

The predicted apparition of comet Halley, at the beginning of the 1980s, provided a way of testing the performance of these new detectors in the field. Provisional magnitude calculations suggested that it might be observable from the beginning of 1982, whilst astrometric calculations could predict its position in the sky to an accuracy better than 1 arcmin. During the first quarter of 1982 the comet was sought without success. This caused some concern. Could the nucleus have broken up? Its predicted trajectory passed behind the Sun during the summer of 1982, hence rendering it unobservable. The search was taken up again in Autumn, with even greater determination. It was a team from Caltech which first announced a sighting, in October 1982. The nucleus, then situated at 11 AU from the Sun, had a visual magnitude of 24, which lay at the very limit of detector sensitivity. It was observed at just 9 arcsec from the position calculated by Don Yeoman's astrometry group (see Fig. 5.4). The observation was made with the Mount Palomar 5 m telescope, using a CCD camera, an exact replica of the one which was to equip the Hubble Space Telescope (HST). One month later, a French team also observed the comet, this time using an electronographic camera, with the 3.6 m telescope at the Canada–France–Hawaii (CFH) Observatory, on Mauna Kea in Hawaii.

From this time on, astronomers tracked the comet relentlessly. The first surprise came in January 1983 when astronomers at the European Southern Observatory (ESO) in Chile were to

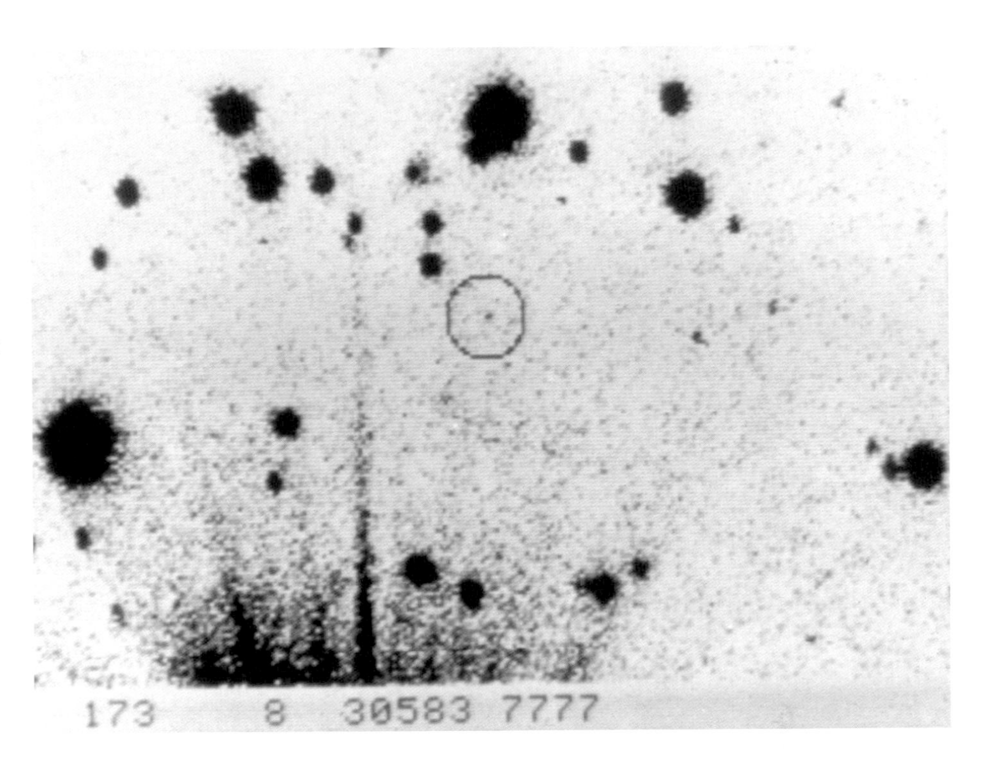

Figure 5.4.
First detection of P/Halley in October 1982. The image was obtained at Mount Palomar, by a team from Caltech using a CCD camera. The comet was detected at a distance of only 9 arcsec from its calculated position. Stars in the field of view are slightly stretched out due to movement of the telescope during exposure. The latter was programmed to compensate for the comet's own motion. The visual magnitude of the comet at the time was only 24. Jewitt and Danielson, 1983, IHW.

observe a five-fold increase in brightness over October observations (see Fig. 5.5). Only a first burst of activity could explain this phenomenon, involving ejection of a pocket of gas, together with the surrounding dust. The apparent surface area of the comet grew over a very short period of time, and the intensity of solar light scattered towards Earth increased in consequence. A month later the comet had attained its expected flux level, corresponding to that of an inactive object. Much can be learned from an unexpected phenomenon of this kind. The comet was still at 8 AU from the Sun and, at this distance, water, assumed to be the principal constituent, could not sublime. The burst which had been observed must have been caused by sublimation of some other volatile molecule.

Encouraged by this first result, observers continued their efforts. Further bursts were observed on several occasions over the next few years, using large Earth-based telescopes, notably on the island of Hawaii and in Chile. By the end of 1986, the comet was observable using 2m-class telescopes (see Fig. 5.6) and, from this time on, amateur astronomers took up the task of monitoring its brightness, thereby playing an essential role in the whole operation.

The spring of 1985 marked the beginning of a new phase. A team from the University of Arizona recorded the first spectrum of the comet, showing emission from the radical CN. It now became possible not only to observe the brightness of the nucleus, but also to make spectroscopic measurements. These could give more information about its activity and help to identify the nature of gases being ejected. In autumn, the first detection of the radical OH was obtained in the radio region, at a wavelength of 18 cm, using the radio telescope at Nançay in France. The daily repetition of this observation over more than a year provided an almost continuous record of the comet's activity.

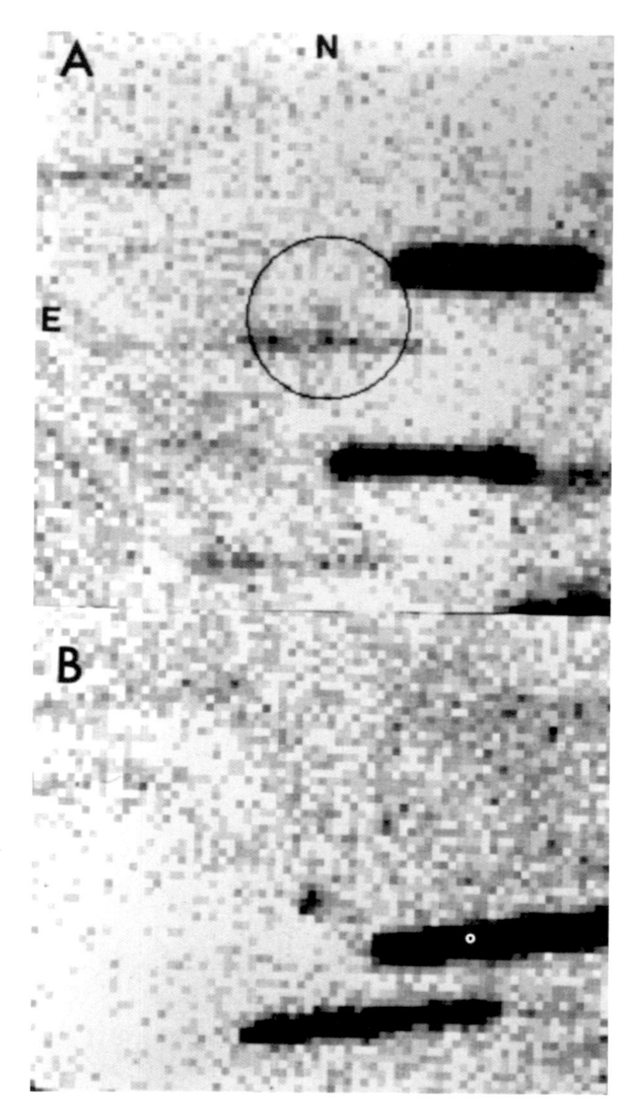

Figure 5.5.
Observations of P/Halley at the ESO on 10 December 1982 (A) and on 14 January 1983 (B). An increase in brightness by one magnitude in the second image is the first evidence of a burst of activity, probably due to sudden release of pockets of gas. Courtesy of ESO.

Complementary OH measurements were carried out in the ultraviolet region by the astronomical satellite International Ultraviolet Explorer (IUE), in orbit around the Earth.

For the 1986 apparition, as mentioned earlier, the geometric configuration of the cometary trajectory was rather unfavourable. The perihelion passage occurred behind the Sun, so that the two most propitious periods for Earth-based observation were November 1985 and April 1986. The northern hemisphere was best disposed for the first of these dates, and the southern hemisphere for the second. The campaign maintained a high level of activity between September 1985 and June 1986, reaching its maximum intensities around these two periods.

In the autumn of 1985, new techniques were introduced and it was only a short time before they bore their first fruits. The HCN molecule was detected for the first time in September 1985. The observation was made using heterodyne spectroscopy with the IRAM 30 m receiver in Grenada, Spain (see Fig. 5.7). This was indeed the first radio observation of a parent molecule directly released from a cometary nucleus (the only other parent molecule known at the time being CO, identified in the UV). It is worth remembering at this point that all previous detections, apart from several rather doubtful results, concerned radicals, atoms or ions produced by dissociation of such parent molecules. In December 1985, an American team achieved another spectacular result: the first direct detection of the molecule H_2O. This event confirmed that water is indeed the principal constituent of the comet. In order to be able to observe cometary water, despite absorption of the relevant frequency bands by Earth's atmosphere, NASA's stratospheric airplane known as the Kuiper Airborne Observatory (KAO) was used to obtain an infrared spectrum of the comet (see Fig. 6.2).

A further remarkable result came from the use of infrared spectroscopy after the comet's perihelion passage, and after its encounter with the various space probes. Following observations

Figure 5.6.
Observation of P/Halley at the Observatoire de Haute-Provence in France on 29 December 1984. The image was obtained using a CCD camera. The arrow shows the position of the comet, whose visual magnitude was then about 21. Courtesy of J. Lecacheux.

carried out with the infrared spectrometer on board the probe Vega 1, many observers attempted to study in more detail an unexpected emission, located between 3 and 4 μm, which had been attributed to hydrocarbons (see Fig. 5.8). These ground-based observations led to a better determination of their general nature, although they could not yet provide a firm identification.

In parallel with these new techniques, traditional imaging and spectroscopic methods continued to be used to good effect in the visible region. Even if the cometary nucleus itself was not accessible to direct observation, jet evolution could be studied by imaging the circumnuclear region. Such data was to prove invaluable in determining the rotation period of the nucleus, and gave rise later to great controversy. Wide

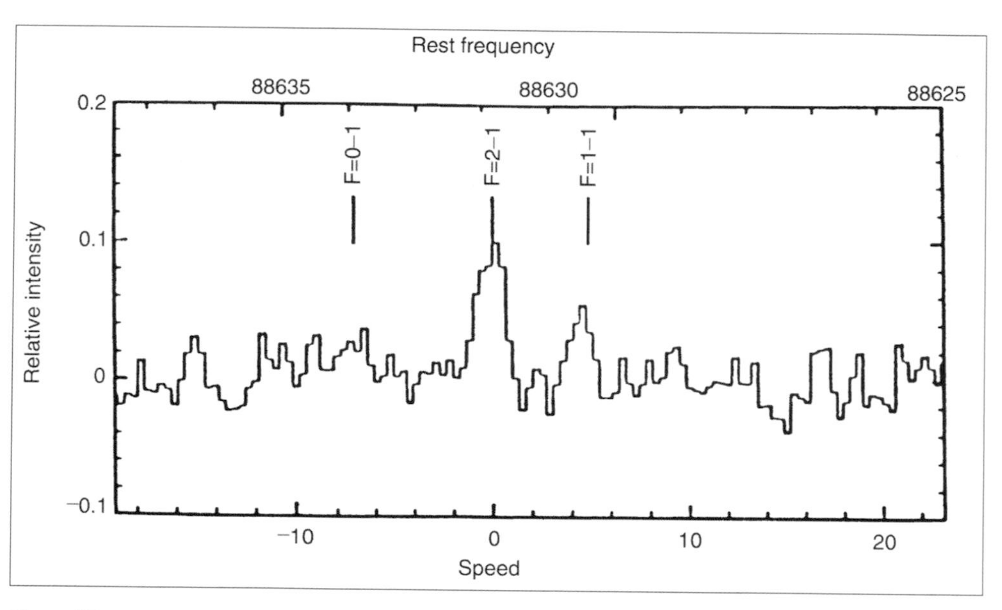

Figure 5.7.
First HCN spectrum for P/Halley, obtained with the 30 m receiver of the Institut de radio-astronomie millimétrique (IRAM) at Pico Veleta in Spain. From Despois *et al.*, 1986 [42].

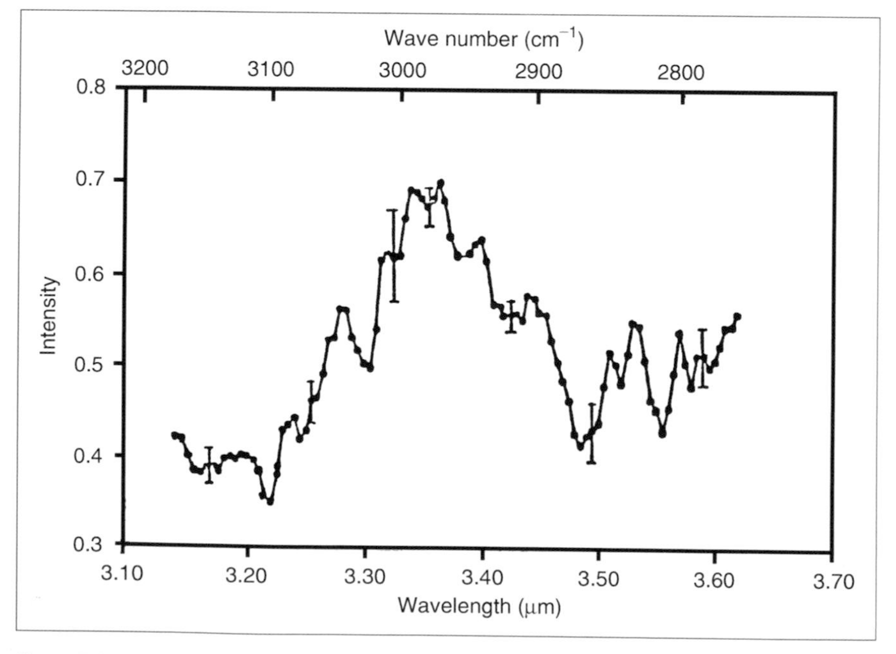

Figure 5.8.
The infrared spectrum of P/Halley from ground-based observations in the range 3.1 to 3.7 μm, using the United Kingdom Infrared Telescope (UKIRT) in Hawaii. From Bass *et al.*, 1982 [26].

field imaging, for its part, was used to study the dust tail, the evolution of the plasma tail as it interacted with the magnetic field in the neighbourhood of the comet, and various phenomena caused by interaction between cometary ions and the solar wind. In addition to this, visible spectroscopy was used to make an accurate map of different radicals (e.g. C_2, CN, NH_2, etc.) in the coma, by spatial analysis along the spectrograph slit. At the end of an intense campaign lasting more than a year, all the results were brought together during two international conferences in autumn 1986, one held in Heidelberg, the other in Paris.

5.3 Space exploration of comet Halley

Five space probes were sent to investigate comet Halley: a European probe, Giotto, two Soviet probes, Vega 1 and Vega 2, and two

Figure 5.9.
The Giotto probe, 1.86 m in diameter and 2.85 m high. At launch, its weight was 960 kg, and it was stabilised by its rotation. The outer walls were covered with solar cells to supply the electrical energy necessary for its operation. Courtesy of ESA.

Japanese probes, Suisei and Sakigake. For the European Space Agency, the Giotto mission took on a particular importance. This was its first mission to go beyond the confines of terrestrial orbit and towards another object in the Solar System. It was a unique opportunity for the new agency to develop a programme of planetary exploration on a par with other major organisations.

For more than twenty years scientists such as the American Fred Whipple and the Belgian Pol Swings had been pleading in favour of a cometary mission. Many reports had been produced on both sides of the Atlantic. Then, in 1978, an ambitious project was announced, which would associate NASA and the ESA, involving a flyby of comet Halley and a follow-up of comet P/Tempel 2. Following NASA's decision to withdraw in January 1980, the ESA decided upon a less ambitious mission, making use of already existing equipment, and devoted solely to a flyby of comet Halley. The Giotto project was approved in July 1980. As the flyby was to take place in March 1986, at the moment when the comet would cross the ecliptic plane for the second time, the schedule for setting up such a project was extremely tight; and, in contrast with other space missions, this deadline could not be extended. Moreover, the aim was to pass as close as possible to the nucleus, within a range of 1000 km, so that the technological challenge facing the ESA was two-fold.

In the USSR, the decision to send a space mission to the comet was taken in January 1981. Like the Europeans, they had to face the challenge of a tight schedule. In this case, the probe was not devoted entirely to the comet itself, but was intended for a prior observation of Venus; after a flyby, it would go on to its encounter with the comet. The Soviets had been developing their Venus exploration programme for many years, with good results. The Venera probes, considerably heavier than Giotto, were equipped with a pointing platform which greatly simplified imaging and spectroscopic measurements. The new project, named Vega (Venus–Halley), comprised two identical probes, and aimed to come within 10 000 km of the comet.

Meanwhile the Japanese, specialists in plasma physics, were particularly interested in interactions between the comet and the solar wind. They thus designed two probes for study of the interplanetary medium, which were sent towards the comet, but at significantly greater distances than the other probes. These also encountered the comet in March 1986.

5.3.1 Space probe instrumentation

Two types of observation are made possible by space exploration of objects in the Solar System. Firstly, techniques used for Earth-based observation, known as *remote sensing*, can be applied in particularly favourable conditions, since the space probe is located relatively close to the body being studied. In this way imaging and spectroscopic techniques, well known to astronomers, can provide detailed studies of the surface and atmosphere of the object. Secondly, the probe may actually penetrate the environment under investigation, where it can carry out other types of measurements, known as *in situ* measurements. This involves the direct analysis of samples taken on board the probe, e.g. counting of particles, mass spectroscopic analysis of grains or gases, analysis of the solar wind, or magnetometry.

The Giotto probe and the two Vega probes, which passed close by the comet, carried out measurements falling into both categories. All three were provided with a camera to photograph the nucleus and its immediate neighbourhood. They also carried several mass

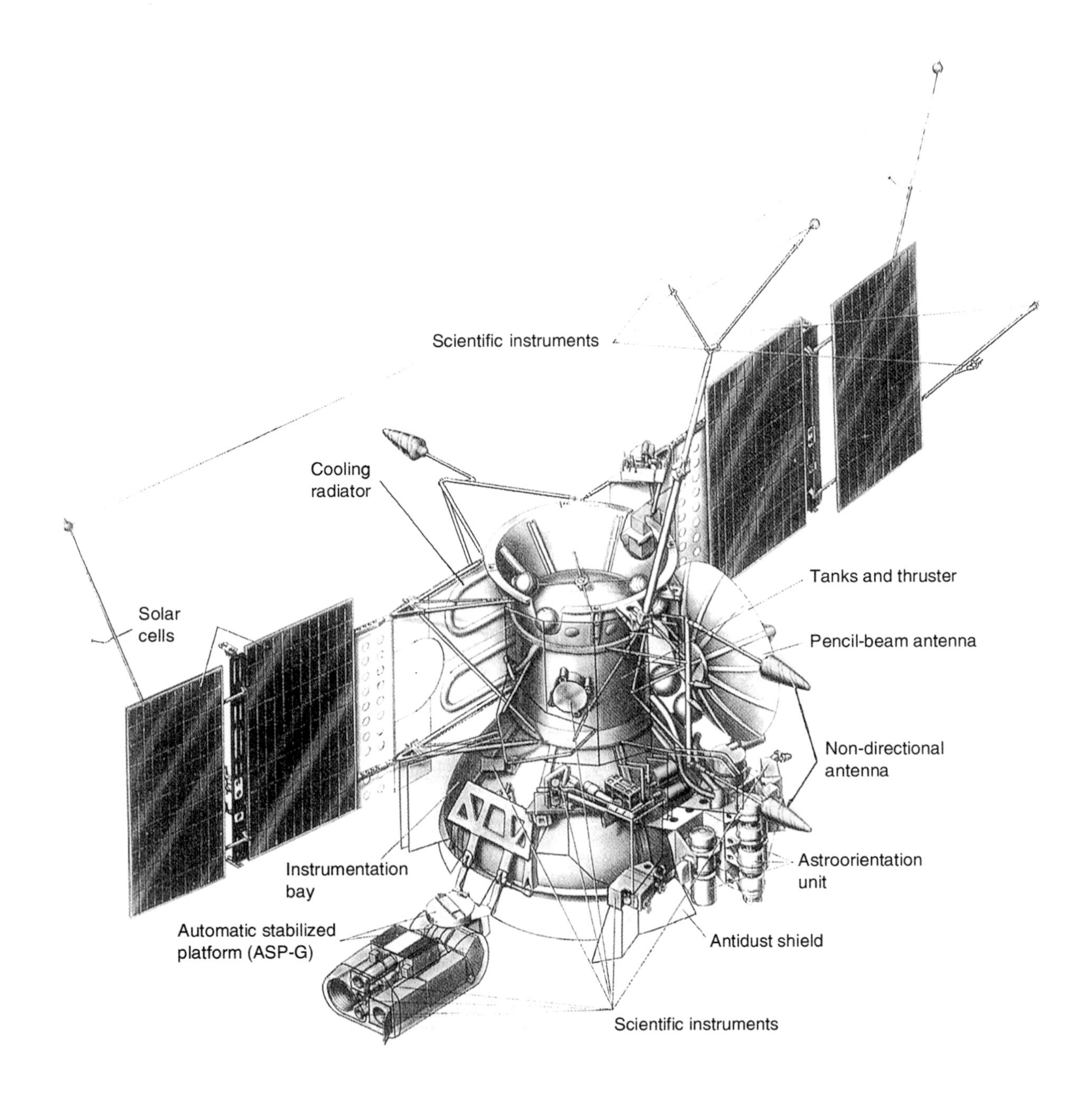

Figure 5.10.
Schematic diagram of the Vega probe. Launched in December 1984, the Soviet probes Vega 1 and Vega 2 were originally directed towards Venus in order to release a balloon in the planet's atmosphere in July 1985. They were then redirected towards comet Halley, which they flew by on 6 and 9 March 1986, respectively. Courtesy of the CNES and the Institute of Space Research, Moscow.

spectrometers, to analyse chemical compositions of neutral gases, ions and cometary grains; and finally, they were equipped with magnetometers and other instruments aimed at studying solar wind particles, radio waves and plasma. In addition, the Giotto probe had a photopolarimeter devoted to studying optical properties of cometary dust particles.

The Vega probes, with their pointing platforms, included two spectrometers operating between the ultraviolet and the infrared. The Japanese probes, which flew by at greater distances from the comet, were designed to study the hydrogen envelope by analysis of its UV radiation in the Lyman α line at 121.5 nm, and also to study the electromagnetic environment of the comet.

Despite considerable complexity in the choice of technology and the close deadline, the space exploration of comet Halley was a great success. Almost all instruments on board Giotto functioned satisfactorily, and all those on the Vega probes were successful on at least one of the two. The Japanese probes also provided the results hoped of them. Even more remarkably, as the Vega probes reached the comet a week before Giotto (on 6 and 9 March 1986), they were able to transmit some final position corrections to the European probe, so that its point of closest approach could be brought down to just 500 km (on 14 March 1986). In scientific terms, the space exploration of comet Halley was a tremendous success, adding a new dimension to cometary physics.

5.4 The results

With his 'dirty snowball' model, suggested in the 1950s, Fred Whipple had implied that the comet was essentially composed of water. The 1986 results proved him right, at last producing unambiguous evidence of water vapour as the main constituent in cometary gases.

5.4.1 The nucleus

Although Whipple's model was thereby confirmed, it was not long before several surprises appeared. It was discovered as soon as flybys were effected that the nucleus was dark, bulky and irregularly shaped, as well as being hotter than had been expected (see Figs. 5.11 and 5.12). A radius of about 3 km had been predicted, and instead, an extended object of rather rectangular dimensions ($8 \times 7 \times 15$ km) was observed. As its size had been underestimated, the assessment of its albedo obtained from measurements of cometary flux had to be revised. Instead of the expected 10 or 20%, the albedo of the nucleus turned out to be only 4%, lower than that of the darkest materials known on the Earth.

A further unexpected discovery was made by analysis of cometary dust grains using mass spectrometers on board the Vega and Giotto probes (see Table 5.1 and Fig. 5.13). This was the abundant presence of 'primitive' grains, that is, grains rich in light elements such as H, C, O and N. These were the most abundant elements in the primitive solar nebula at the time when the Solar System was formed. Grains composed of complex hydrocarbons appear to be particularly abundant. To this was added the discovery by the infrared spectrometer IKS-Vega of an emission band at a wavelength around 3 μm, corresponding to hydrocarbons in either gaseous or solid state. Similar spectral signatures had already been noticed in other contexts, namely in certain interstellar spectra and also in laboratory produced spectra of ices which had been subjected to intense UV radiation or irradiation by energetic particles.

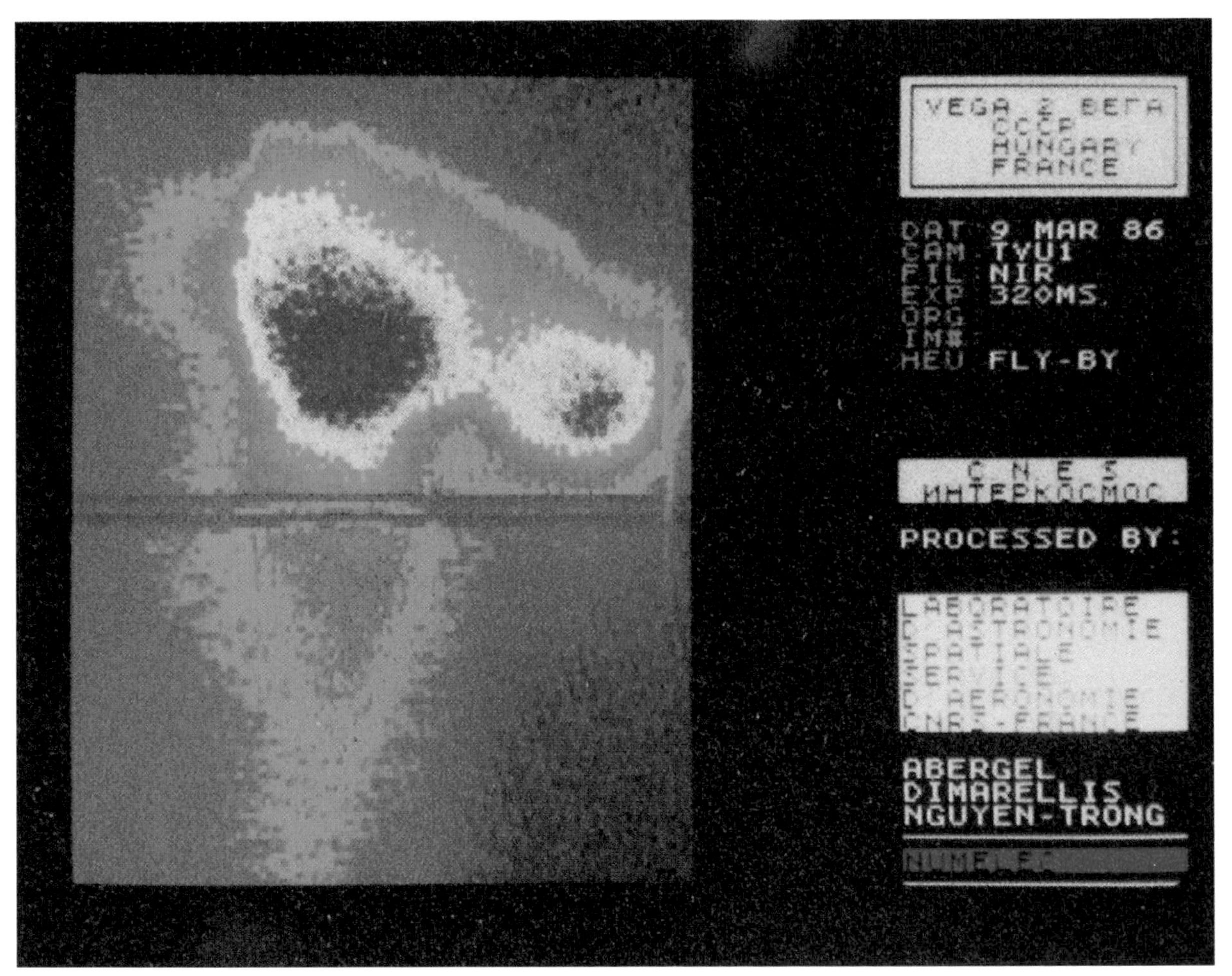

Figure 5.11.
Image of the Halley nucleus, taken by the camera on board Vega 2 when it was at a distance of about 8000 km. It can be considered as a landmark, obtained almost in real time and without sophisticated image processing. The irregular structure of the central region of the comet is revealed. Courtesy of CNES and the Institute of Space Research, Moscow.

A new idea emerges from an overall view of these results. The cometary nucleus may be at least partially covered by a refractory organic layer, presumably produced by the effects of solar ultraviolet radiation and cosmic rays on organic ices. This hypothesis implies a strong interaction between cometary matter and the interstellar medium, whilst maintaining Whipple's model intact. It would appear, however, that the snowball really is very dirty!

Apart from its extremely low albedo, the cometary nucleus exhibits another quite remarkable feature. It is not at all homogeneous. Indeed, there seem to be active regions from which gas and dust escape, and inactive regions of higher temperature. It was expected that active regions would display temperatures close to the sublimation temperature of the ice, around 200 or 220 K. Inactive regions, on the other hand, must be in equilibrium with the

Figure 5.12.
The nucleus of comet Halley, photographed by the camera on Giotto when it was at a distance of about 18 000 km. The nucleus is very dark and irregularly shaped ($15 \times 7.5 \times 7.5$ km). Photo courtesy of ESA/Max Planck Institut für Aeronomie, with the kind permission of H.U. Keller.

solar radiation they receive, implying a temperature of about 300 K for a heliocentric distance of 1 AU. It was this last value which was measured by the Vega infrared instrument. The result was comparable with earlier ground-based observations which had shown a high temperature for the nuclei of certain old comets. An idea was gradually emerging according to which the inactive part of a comet's nucleus might increase relative to its active part as the comet aged, whilst the ices contained within it would sublime at successive passages close to the Sun. If this is accurate, the traditional dividing line between asteroids and comets might well be more difficult to draw than had previously been supposed. Certain asteroids might just be old comets.

However, many questions remain unanswered. Does this organic material completely coat the nucleus and if so, how thick is it? We know that at each perihelion passage the comet loses about the equivalent of one metre thickness of matter. Can the coating resist the consequences of perihelion passage? If it is produced by irradiation, how does it reconstitute itself in such a short lapse of time (on the astronomical scale) as just 76 years? Or should the organic material be considered as intimately bound up with ices in the interior regions of the comet, a result of some massive irradiation occurring in the early stages of the Solar System, before the cometary nucleus had formed? In order to obtain an answer, not only the surface, but even the interior region of the nucleus will have to be the subject of *in situ*

Table 5.1. Atomic abundances in comet Halley, the Sun and meteorites.

Atom	Sun[a]	Meteorites[b]	P/Halley[c]	
			Dust alone	Dust and ice
H	2.6×10^6	490	2000	4060
C	940	70	810	1010
N	260	5.6	42	95
O	2200	712	890	2040
Na	5.3	5.3	10	10
Mg	100	100	100	100
Al	7.9	7.9	6.8	6.8
Si	93	93	180	180
S	48	48	72	72
K	0.35	0.35	0.2	0.2
Ca	5.7	5.7	6.3	6.3
Ti	0.22	0.22	0.4	0.4
Cr	1.3	1.3	0.9	0.9
Mn	0.88	0.88	0.5	0.5
Fe	84	84	52	52
Co	0.21	0.21	0.3	0.3
Ni	4.6	4.6	4.1	4.1

Abundances normalised with respect to magnesium.
[a]Based on photospheric spectroscopy.
[b]Based on analysis of a primitive meteorite (a carbonaceous chondrite).
[c]Based on analysis of dust from P/Halley, using the Vega mass spectrometer, and on the ice composition deduced from spectroscopy of volatiles in the coma.

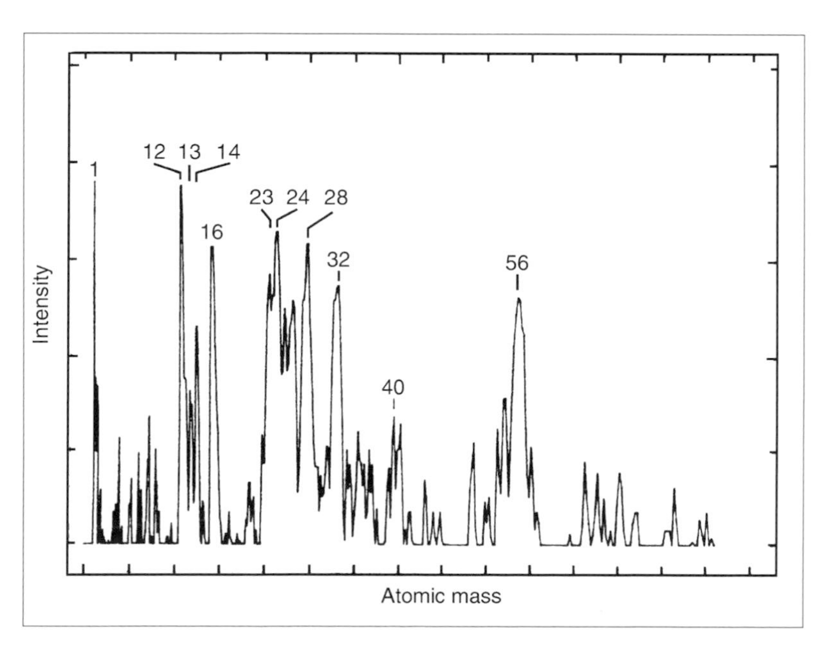

Figure 5.13.
Composition analysis of a cometary grain made by mass spectrometry on board the Vega 1 probe. The spectrum is dominated by light elements, whose signatures appear at atomic masses 1 (H), 12 (C), 14 (N) and 16 (O). From Kissel and Sagdeev, 1986 [57].

analysis. This is indeed the aim of the European mission Rosetta (see Chapter 11).

Yet another debate was raging in the months following the Halley campaign, this time concerning the spin period of the nucleus. The question had already arisen many decades before, but despite considerable effort on the part of astronomers, this particular parameter had eluded all attempts at determination from the wealth of images produced in 1910. Repeated photometric observations made between 1982 and 1986 proved no more able to resolve the difficulty. It was only known that rotation was slow, of period greater than one day. The space-based observations being too brief for application to this question, all hopes rested with ground-based monitoring during 1986. However, observational results proved to be contradictory, and a controversy arose between the partisans of a two day period and those who preferred a seven day period. Lacking any universally acceptable solution, a two-component model was proposed in which a two day rotation was superposed upon a slow precession (greater than seven days). Other examples of comets with complex rotations have subsequently been found (see Table 7.1).

5.4.2 The coma

It was the nucleus of comet Halley which attracted most attention, and justifiably so, after the flyby of the various space probes; for it had remained totally unknown until then. However, it would be a mistake to underestimate the significance of space and terrestrial observations relating to dust and gas in the coma.

Apart from confirming the presence of water vapour, the main constituent of the comet, several new molecules were identified. HCN was detected for the first time, by ground-based observation in millimetre wavelengths. Carbon dioxide, CO_2, was identified for the first time, using the infrared spectrometer IKS-Vega, and its abundance was confirmed by mass spectrometry. Carbon monoxide, CO, already observed on many comets, was once again detected by ultraviolet and infrared spectrometry, as well as by mass spectrometry. Other molecules were identified in much lesser quantities by the IKS spectrometer: formaldehyde, H_2CO, and possibly OCS (see Fig. 5.14). Remarkably, some expected molecules such as methane, CH_4, and ammonia, NH_3, were not directly detected. Following a more detailed analysis of mass spectrometry experiments, extremely low mixing ratios, of the order of 1% relative to H_2O, could be deduced.

As mentioned earlier, complex hydrocarbons were detected by their infrared spectral signature, although the exact nature of these molecules remains uncertain. It has been suggested that some of them may be comprised of polyaromatic hydrocarbons (PAH), molecules already well known to astrophysicists, who believe they may have detected them on many occasions in the interstellar medium. They are also well known to laboratory chemists, for they are commonly found on Earth, notably in soot. The TKS-Vega spectrometer, operating in the ultraviolet, would seem to have revealed the spectral signature of two such PAHs, namely naphthalene, $C_{10}H_8$, and phenanthrene, $C_{14}H_{10}$. This discovery brings out once again the analogy between cometary material and the interstellar medium (see Fig. 5.15). However, it could not be confirmed in recent observations of comets Hyakutake and Hale–Bopp.

A further clue to the presence of complex hydrocarbons in the coma was revealed by one of the mass spectrometers on board a space probe. This was a periodic emission, appearing at atomic masses separated by 30 units. It has been

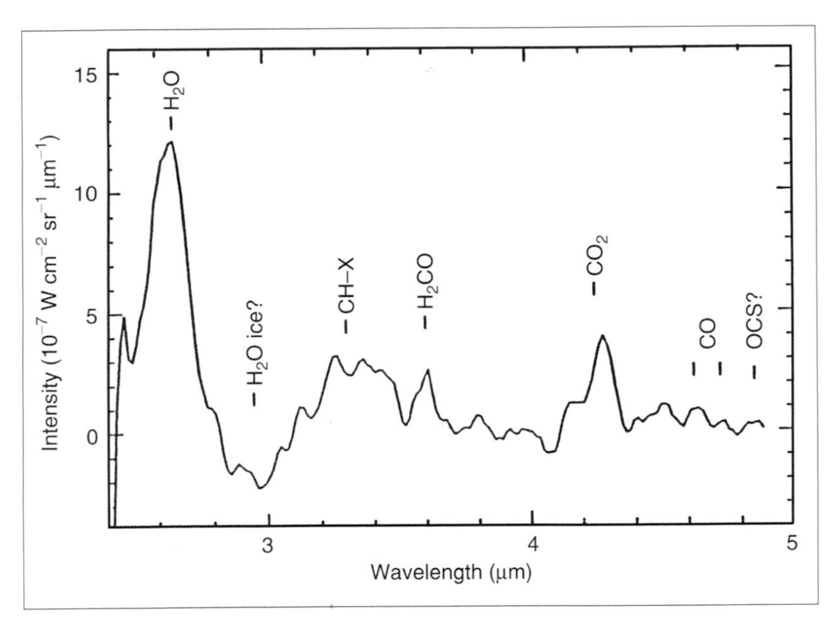

Figure 5.14.
Infrared spectrum of comet Halley, between 2.5 and 5 μm, registered by the IKS spectrometer on board Vega 1. Signatures of H_2O, CO_2 and CO are visible, as is an emission due to hydrocarbon molecules and/or grains. Other molecules may also be present (e.g. H_2CO, OCS). From Combes *et al.*, 1988 [35].

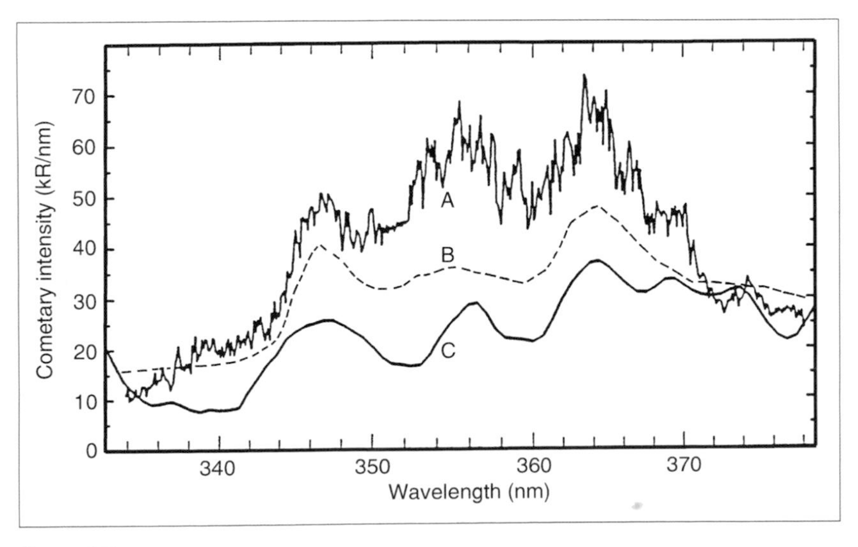

Figure 5.15.
Ultraviolet emission attributed to PAHs, observed by TKS-Vega. A comparison is made between the spectrum of the inner coma (C) and two laboratory spectra of phenanthrene, one being the spectrum of a molecular jet excited by laser (A), the other being for a solution in cyclohexane (B). From Moreels *et al.*, 1994 [66].

proposed that these signatures may have been caused by polymers of formaldehyde, viz. polyoxymethylene $(H_2CO)_n$. More generally, it would seem that they could result from a wide range of hydrocarbons, too complex for unambiguous identification.

5.4.3 Relative isotopic abundances

Relative abundances of isotopes were also measured as part of the exploration of comet Halley. Such observations are important, for these isotopic ratios give information about conditions prevailing when the molecules containing the isotopes were formed.

A relatively high proportion of deuterium was observed by mass spectroscopy (see Fig. 9.4). This result should be compared with deuterium excesses measured in certain interstellar molecules, such as HCN and HNC. This excess can be explained by a fractionation effect resulting from reactions between molecules and ions occurring in the interstellar medium. The deuterium excess in the comet based upon measurements of HDO would seem to suggest that H_2O molecules trapped in cometary ices have been subjected to the same enrichment processes, in the depths of the interstellar medium, before being incorporated into the nucleus of comet Halley.

This new information has profoundly altered our conception of the cometary nucleus. As will be further discussed in Chapter 9, analogies with the interstellar medium are becoming more relevant: all cometary molecules have been previously detected in the interstellar medium. Furthermore, in both these environments, gaseous carbon occurs rather in the form of CO than as CH_4, and ammonia is relatively scarce. In both cases, part of the carbon not assimilated into CO may well be occurring in the form of macromolecules of PAH-type. Another part would appear to be assimilated into refractory organic materials produced by the effects on organic ices of intense ultraviolet radiation or irradiation by energetic particles. Likewise, measurements of relative isotopic abundances show a clear analogy between cometary material and certain 'primitive' meteorites.

5.5 Space exploration of P/Grigg–Skjellerup

The Halley flyby did not mark the end of Giotto's exploits. Several years later, the European probe had been subjected to a series of tests and redirected towards a new, less active comet, 26P/Grigg–Skjellerup. The mission was named Giotto Extended Mission (GEM) and the flyby took place on 10 July 1992.

Some of the scientific instruments on board Giotto had suffered in the close encounter with comet Halley. This had involved many collisions with cometary dust grains. In particular, the camera and the mass spectrometers were no longer operational. But several other instruments, notably the optical photopolarimeter and plasma physics experiments, gave good results, thereby allowing an interesting comparative study to be made of two comets exhibiting very different levels of activity.

The Giotto probe came within 200 km of the Grigg–Skjellerup nucleus, with flyby on the antisolar side of the comet. Before flyby, the photopolarimeter began to record a signal from cometary dust at a distance of 17 000 km. As in the case of comet Halley, the time development of this signal could be used to determine optical properties of cometary grains, and also to reveal some differences between the inner coma, dust jets and outer coma.

5.6 Ionised gases and solar wind interactions

As mentioned earlier, it was around the end of the 1950s that Earth-based wide-field imaging observations of comets led L. Biermann and H. Alfvén to propose an interaction model in which the lines of magnetic field fold around either side of the nucleus. Several scenarios could be envisaged, depending on the level of activity of the comet. For a low level of activity, the solar wind would pass directly around the inert nucleus; but in the case of an active nucleus, a bow shock is created, across which plasma is compressed and decelerated, and it is this which deviates flow around the obstacle. The region lying between bow shock and nucleus is the *magnetosheath*, in which both plasma and magnetic field are compressed. In the case of an extremely active comet, the gaseous envelope created by sublimation of nuclear ices may undergo a supersonic expansion before going into a subsonic flow beyond an inner bow shock. In any of these cases, a folding of the magnetic field lines would be expected on either side of the nucleus, a situation which would

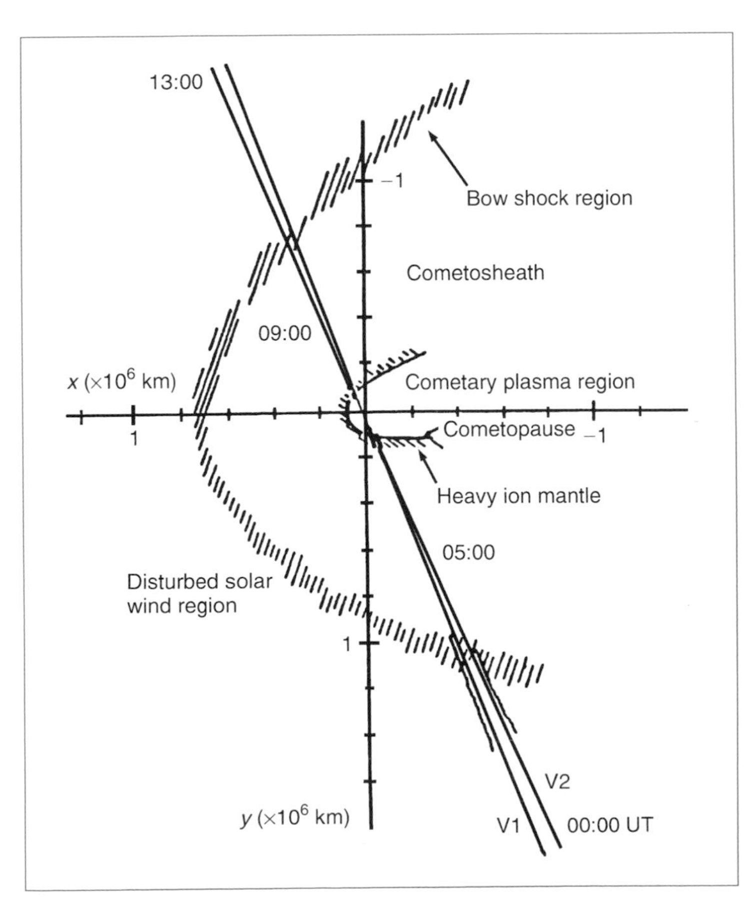

Figure 5.16.
Environment of comet Halley, based on observations made by the Vega probes, displaying the bow shock, cometosheath and cometopause. From Gringauz *et al.*, 1986 [49].

induce lobes of opposite polarity, separated by a high density *neutral sheet* where currents would go to zero.

These ideas have been put to the test by space exploration of the three comets Giacobini–Zinner in 1985, Halley in 1986 and Grigg–Skjellerup in 1992. Although the general outline of Biermann and Alfvén's scenario was confirmed, the actual situation turns out to be rather more complex than their theory had led people to expect.

5.6.1 The ICE probe and comet Giacobini–Zinner

At the end of 1980, NASA hurriedly organised another cometary mission using the satellite ISEE-3, renamed the International Cometary Explorer (ICE) for the purpose, redirected towards comet 21P/Giacobini–Zinner.

Instrumentation on board the ISEE-3 probe consisted mainly of particle analysers together with radio wave and plasma detectors, which were aimed at study of the interplanetary medium. Consequently, exploration of comet Giacobini–Zinner was restricted to study of its interaction with the solar wind, without direct analysis of the comet itself.

The ICE probe flew by comet Giacobini–Zinner on 11 September 1985, with a closest approach distance of 7800 km. Its trajectory during flyby was such that it passed on the antisolar side of the comet, thereby crossing the plasma tail, and was thus well placed to seek out the predicted neutral sheet.

At a distance of several million kilometres from the comet, the probe began to identify high energy particles, with energies lying between several tens of keV and a few MeV. It found some evidence for a bow shock at a distance of about one million kilometres from the nucleus, and detected turbulence from a distance of about 150 000 km down to 20 000 km. Closer still, it crossed the region of the two lobes in which the magnetic field folds around the nucleus. The intensity of this field was of the order of 60 nT, and in opposite directions on either side of the neutral sheet, in which it fell almost to zero. Within the lobes, the density varied inversely with the square of the cometocentric distance, but increased rapidly in the neutral sheet to a value of 1000 electrons per cm^3. The model suggested by Biermann and Alfvén was thus corroborated.

The ICE probe also detected ions in the atomic mass range between 14 and 33. The most abundant were water group ions H_2O^+, but an emission group also appeared at atomic masses 23 and 24, which was attributed to Na^+ and Mg^+. CO^+ was likewise shown to be present.

5.6.2 Space exploration of the Halley ion environment

The solar wind interactions of comet Halley were investigated by the five space probes, all equipped with suitable instruments (radio wave detectors, particle analysers and magnetometers). However, in contrast to the ICE mission, these five probes passed between the Sun and the cometary nucleus, and could not therefore study the neutral sheet. In the case of the Giotto and Vega probes, the choice of trajectory had been imposed by other scientific objectives, the priority being observation of the nucleus, so that it was essential to view the face illuminated by the Sun.

Observations made by the Giotto and Vega probes confirmed the global structure proposed for the comet's solar wind interactions and showed the presence of a bow shock. However, they produced a much more complex image than the one expected. The bow shock was indeed detected, at around one million kilometres from the nucleus. However, upstream, in the direction of the Sun, cometary ions were found implanted in the solar wind throughout a region stretching for a further several million kilometres. A pre-shock region was also recorded, slightly upstream of the main bow shock. Downstream of the bow shock, the magnetosheath, or *cometosheath*, extended as far as the ionopause at a distance of 4000 km from the nucleus. Beyond this, the ionosphere appears as a cavity in which the magnetic field is zero.

The cometosheath can be divided into three distinct regions: a turbulent outer region, a 'mysterious' central region, and a calm inner region. The so-called 'mystery' region has a thickness of around 20 000 or 30 000 km and corresponds to the region in which the nucleon density of cometary water group ions balances the proton density of the solar wind. Note that this mystery region was also detected in the neighbourhood of comet Giacobini–Zinner by the ICE probe. A further boundary is encountered within the calm region, at about 130 000 km from the nucleus. This is the cometopause, a region of magnetic pile-up characterised by a sudden decrease in electron density. Any closer to the nucleus, the contaminated solar wind is rapidly decelerated by collision with neutral elements originating in the nucleus.

Another important result of plasma physics experiments was the discovery by mass spectrometry of many heavy ions. The most abundant were water group ions, clustered around mass 18, but CO, S and also CO_2 groups were likewise detected. Above 50 atomic mass units (amu), the spectra exhibit a series of peaks separated by about 15 amu. Such a signature characterises molecules comprised of C, H, O and N, with a C:O ratio lying in the range 1.2 to 1.4, and a nitrogen abundance below 8%. As indicated above, such organic compounds were brought to light in other experiments. They may have been produced through irradiation of cometary ices by intense ultraviolet radiation or a flux of high energy particles.

In addition, negative ions were detected outside the terrestrial ionosphere for the first time. They were observed by Giotto in the inner coma of comet Halley, and had atomic masses in the neighbourhood of 17, 30 and 100. The corresponding ions are probably O^-, OH^-, C^-, CH^-, NC^-, and ions of complex hydrocarbons. These ions occurred with densities 100 times greater than expected, and the discrepancy with theoretical accounts is still not well understood.

5.6.3 Ion environment of comet Grigg–Skjellerup

Compared with Giotto's first target, comet Grigg–Skjellerup manifests a relatively low level of activity. Its production rate is about a hundred times lower and its solar wind interaction extends over a correspondingly reduced region, viz. around 60 000 km long, compared with Halley's 2 000 000 km. Despite this difference of scale, the structure of the interaction is remarkably similar in the two cases. In particular, there is a bow shock with a pre-shock region just upstream, the cometosheath has a 'mystery' region, and there is a cometopause. Bearing in mind the different levels of activity of the two comets, it is reasonable to assume that the schema originally proposed by Biermann and Alfvén, in general agreement with the structure observed for comet Halley, provides a satisfactory description of cometary ion environments.

6 New techniques: infrared, radio and X-ray

We have already seen that exploration of P/Halley involved other techniques than visible imaging and visible and UV spectroscopy. Infrared and radio observations were also carried out. These latter techniques underwent tremendous advances during exploration of P/Halley, although they had already been in existence for several decades. Moreover, they were developed even further during studies of bright comets that appeared after P/Halley.

Wilson 1987 VII was a new comet discovered more than six months before its perihelion passage. This allowed sufficient time for observations to be planned. With a period of 70 years, P/Brorsen–Metcalf 1989 X is a Halley-type comet which, although fainter, made a favourable return in 1989. Austin 1990 V was less bright than had originally been predicted, but its passage close to Earth (at just 0.24 AU in May 1990) allowed some interesting observations to be made. Levy 1990 XX was very bright in August and September 1990, once again due to a close passage, at only 0.43 AU from Earth. P/Swift–Tuttle 1992 XXVIII is yet another Halley-type comet, which had only once been observed, at its previous passage in 1862. Because its orbit was not accurately known, it was not located until three months after its perihelion passage on 12 December 1992. The dust trail spread out along its trajectory actually intersects the Earth's orbit, and causes the Perseid meteor shower, observable every month of August.

But the most significant progress has come towards the end of the twentieth century from observations of two exceptional comets: C/1996 B2 (Hyakutake) and C/1995 O1 (Hale–Bopp).

Comet C/1996 B2 (Hyakutake) was discovered on 30 January 1996 by the Japanese amateur Yuji Hyakutake. It was soon recognized that this comet was to make a close passage, at only 0.10 AU from Earth, on 24 March 1996. This is one of the closest approaches a comet has made to Earth this century, after 7/Pons–Winnecke (0.039 AU in 1927), IRAS–Araki–Alcock 1983 VII (0.031 AU) and Sugano–Saigusa–Fujikawa 1983 V (0.063 AU). Going back further, there have been no closer passages recorded since the eighteenth century when comet Lexell 1770 I passed at only 0.015 AU in 1770. Comet Hyakutake was a highly productive comet with a gas production rate of about 2×10^{29} molecules per second at the moment of its closest approach to Earth. It then reached a total visual magnitude of about 0, making it a very spectacular object to all (even inexperienced) observers. Its gas production rate increased further as it approached perihelion at 0.23 AU from the Sun on 1 May

1996. At that time, however, the comet was rapidly receding from Earth and observing conditions were worsening. With an orbital period of about 9000 years, this comet belongs to the class of *new comets* recently escaped from the Oort cloud. Although notice was short, a significant observing campaign was nevertheless set up. It should be mentioned that the comet's clear presence in the sky was a very persuasive argument in convincing our colleagues to change long-standing observation projects at large telescopes into last minute cometary observations!

Observations of comet C/1995 O1 (Hale–Bopp) follow quite a different story. The comet was discovered at the end of July 1995 by American amateurs Alan Hale and Thomas Bopp. Located 7 AU from the Sun at the time of its discovery, it was already very bright, at mag 11. This is more than 100 times brighter than comet Halley at the same distance. Prediscovery observations retrieved from archived photographic plates revealed that the comet was already active at 13.1 AU and perhaps even 16.7 AU from the Sun. The comet was expected to pass perihelion at 0.9 AU from the Sun, on 1 April 1997. This gave ample time to plan observations, using most of the major astronomical facilities. In particular, it was possible to monitor the evolution of its activity in detail over a large range of heliocentric distances. Predictions that this comet might become exceptionally bright were borne out when it reached mag −2 around perihelion and was visible to the unaided eye for several months. Its gas production rate (10^{31} molecules per second) was then ten times greater than that of comet Halley at perihelion. This made it one of the most spectacular comets on historical record and gave an opportunity to perform unprecedented observations with modern instrumentation. Like comet Hyakutake, Hale–Bopp is a long-period comet (its orbital period of 4200 years changed to 2400 years after perturbation by Jupiter in 1996), presumably released from the Oort cloud.

6.1 Infrared observations

6.1.1 The beginnings

Comets are relatively cold bodies. Their equilibrium surface temperature is around 300 K at 1 AU from the Sun (for those parts not covered by ices) and about 150 K at 4 AU. In consequence, their nuclei and dust grains, which emit in a first approximation like black bodies, exhibit an emission maximum at wavelength 10 μm when located at 1 AU, and at wavelength 20 μm when located at 4 AU. Hence the great value in carrying out infrared observations of these bodies. However, such observations are not easy, for several reasons. Firstly, the Earth's atmosphere is opaque to a large range of infrared wavelengths, leaving open only a few so-called windows favourable to observation (see Fig. 4.2). Atmospheric molecules such as CO, CO_2, H_2O, CH_4, and others block infrared radiation, and this leads to the well-known greenhouse effect which controls the climate on Earth. In addition to this, the environment of the observatory – the telescope, the dome, and the ground – is at a temperature of about 300 K, and will itself exhibit an emission maximum at wavelengths around 10 μm. Astronomers must be able to distinguish the faint radiation of bodies they wish to observe from this enormous ambient radiation. Moreover, high-performance infrared detectors have only been available for a few years. The first infrared observations of comets were wide band photometric measurements (see Fig. 6.1) carried out on comets Ikeya–Seki 1965 VIII, Bennett 1970 II, and Kohoutek 1973 XII.

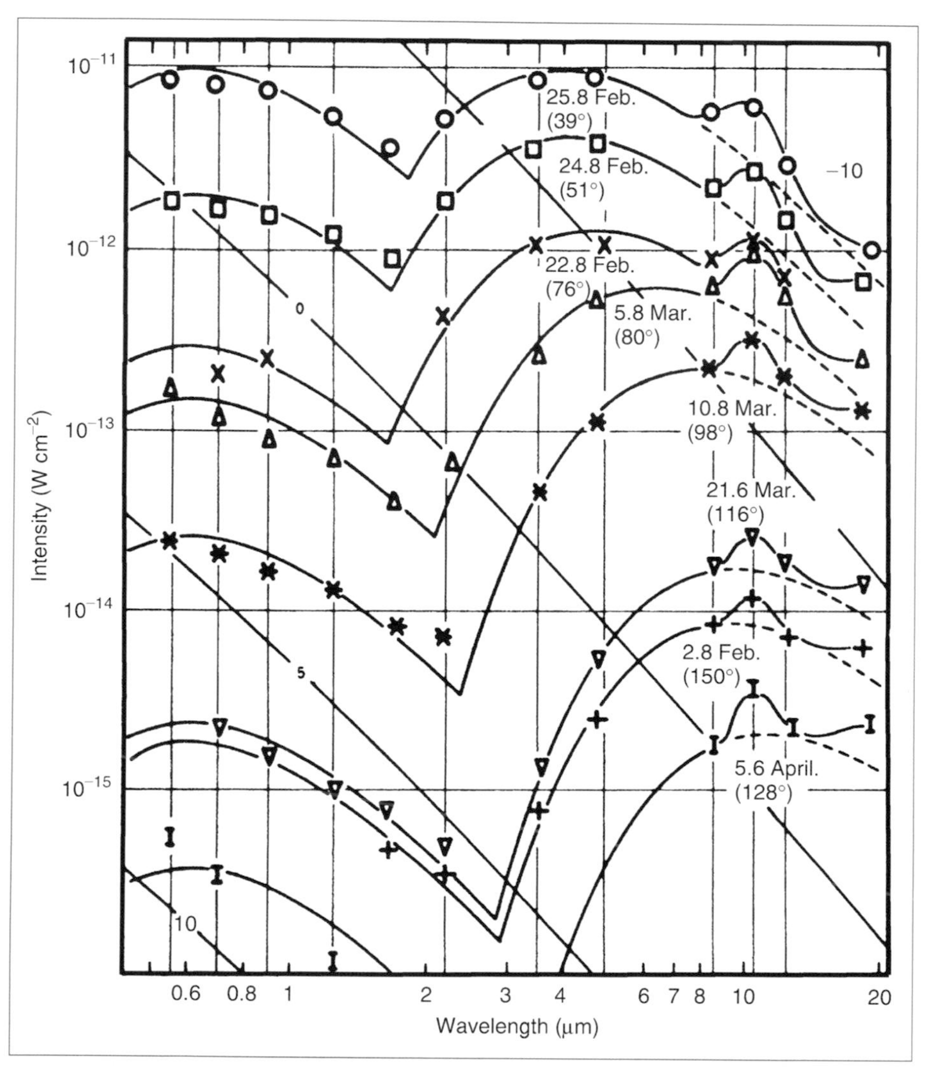

Figure 6.1.
The *continuum* infrared spectrum of comet West 1976 VI as a function of its heliocentric distance. Observe the reflected solar continuum at short wavelengths, and thermal emission at long wavelength. The maximum thermal emission moves to shorter wavelengths as the comet approaches the Sun and its dusts are heated. An emission excess at about 10 μm is attributed to the presence of silicate grains. Taken from Ney, 1982 [70].

These provided measurements of dust temperature and the way it varied with heliocentric distance. Deviations from the blackbody radiation curve were observed; in particular, emission excesses at 12 and 18 μm, attributed to the mineralogical nature of cometary grains, which are rich in silicates such as olivine and pyroxene.

6.1.2 Observations by IRAS

The satellite IRAS (InfraRed Astronomical Satellite), in operation from January to November in 1983, was the result of cooperation between the United States (NASA), Great Britain and the Netherlands. It carried a telescope of diameter 60 cm, cooled by liquid helium in order to reduce background radiation.

Detector arrays operating at wavelengths 12, 25, 60 and 100 μm made it an ideal instrument for mapping the infrared sky. During a systematic exploration of the sky, IRAS detected thirty comets, among which six new ones carry the name IRAS, sometimes together with the names of other discoverers.

One particularly interesting comet discovered by this satellite was IRAS–Araki–Alcock 1983 VII. This intrinsically faint comet would have remained altogether unremarkable if it had not passed at just 0.031 AU from Earth on 11 May 1983. Observations had to be organised very rapidly, for the comet was discovered only 16 days before this extremely close encounter. It was not possible to resolve the nucleus, but it remains one of just two comets in which the molecule S_2 has been detected by its ultraviolet bands (see Chapter 4). This molecule has a very short lifetime and stays close to the nucleus.

One of the great discoveries made by IRAS was that of *cometary trails*, narrow bands of dust following the orbits of certain short-period comets. IRAS was able to identify such trails for P/Tempel 2, P/Encke, P/Gunn, P/Kopff and P/Schwassmann–Wachmann 1. They should not be confused with tails, which always follow the anti-solar direction, whereas cometary trails lie along the cometary trajectory, both upstream and downstream of the comet. They are not apparent on visible images and, as yet, have only been revealed in the infrared. IRAS has also discovered several trails with which no comet can be associated.

Cometary trails are due to large dust grains, which are largely insensitive to solar radiation pressure and hence unlikely to form classic dust tails. They would be ejected from the nucleus at low speeds and gradually spread out along the cometary orbit. Such grains are very likely the same as those causing meteor showers, observed from Earth when it crosses a cometary orbit.

The destiny of a cometary dust grain thus depends very much on its size. The largest are of the order of 1 cm, since larger grains could not detach themselves from the cometary nucleus. These would remain close to the nucleus for long periods, forming an almost permanent coma which might even give the impression of enduring activity. They are responsible for most of the radar echoes and *continuum* radio emission. Smaller grains, of the order of 1 mm, would drift slowly along the orbit to form cometary trails and, in certain cases, meteor showers, whilst still smaller grains would constitute the tails. Only the latter appear to us in visible images.

6.1.3 Infrared spectroscopy and parent molecules

Infrared spectroscopy is an excellent technique for identifying and studying parent molecules which originate in the ices of cometary nuclei (see Section 4.3). The first few incursions into the near infrared (at around 1 μm), revealed nothing but radicals, just as for the visible region. Infrared molecular spectroscopy was first seriously applied to comets during the Halley campaign.

The first result was detection, at long last, of the water molecule. Since the Earth's atmosphere is opaque to H_2O bands, this observation was made from a stratospheric plane, the KAO (Kuiper Airborne Observatory, now decommissioned), specially equipped for astronomical observation by NASA. At the altitude from which the observations were made (13 000 m), absorption by telluric lines of water remains significant. However, these lines are narrow and could be revealed by making observations at times when the comet had a large radial velocity relative to Earth (of the order of 20 km/s), so that the cometary lines were Doppler shifted relative to terrestrial absorption lines. Such observations were successfully made for comets P/Halley and Wilson 1987 VII (see Fig. 6.2).

The presence of water was thereby demonstrated, and its abundance was in good agreement with values obtained indirectly. This came as no surprise, but in addition, these observations were made with high enough spectral resolution to be able to measure individual rotational-vibrational lines. The distribution of molecules over their various energy levels was thus established in detail, and their temperature determined.

In situ observations of P/Halley's infrared spectrum, made by the Vega probes, are discussed in Chapter 5. However, let us reconsider the strong emission these probes revealed around 3.4 μm. This spectral region can be observed from Earth, and the emission was confirmed shortly afterwards by further observations of P/Halley and of all the other bright comets (half a dozen or so) which have appeared since. Such emission is characteristic of the CH group, occurring in all organic molecules (hydrocarbons, alcohols,

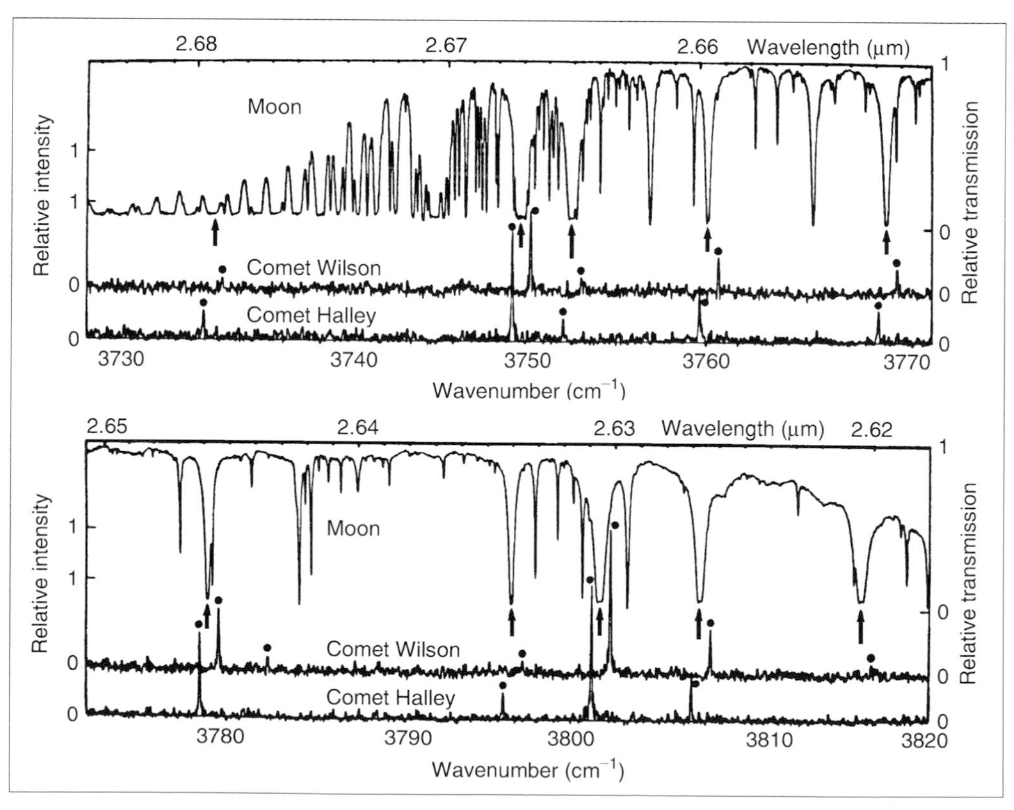

Figure 6.2.
H_2O spectra observed for comet P/Halley on 26 March 1986 and comet Wilson on 12 April 1987, using the stratospheric plane KAO. The spectrum for the Moon has been reproduced above the cometary spectra to show atmospheric transmission. Cometary water bands, marked by *large dots*, are shifted relative to absorption lines of atmospheric water, marked by *arrows*. This is due to the Doppler effect; velocities relative to Earth were –47 km/s for comet Wilson and +35 km/s for Halley. Taken from Larson *et al.*, 1989 [60].

ethers, aldehydes, etc.), but its exact origin has remained a mystery for many years. The problem is whether emission results from fluorescence of small molecules, infrared fluorescence of large molecules excited by solar UV radiation, or from small heated grains with an organic coating. Further information must be obtained, either by observing this region with high spectral resolution, or by observing other wavelength regions in the radio or infrared, before we can determine the exact nature of those molecules causing this emission.

A determining factor in this identification was the discovery in 1990 of cometary methanol (CH_3OH) through its radio lines (see later). Observations implied a significant abundance for the molecule. It also has strong vibrational bands around 3.3 μm, one of which is centred on 3.52 μm and well separated from its neighbours. It can be clearly seen in the better known cometary spectra, such as the spectrum of comet Levy 1990 XX (see Fig. 6.3). Other bands, lying around 3.4 μm, must contribute to the signal observed at these wavelengths. If the expected signal from methanol is calculated on the basis of the isolated band at 3.52 μm or, when possible, from the methanol abundance as measured by its radio emission, it is found that the presence of methanol cannot alone explain all the emission observed around 3.3 μm. As shown in Fig. 6.3, an excess emission remains around 3.4 μm, together with an isolated band at 3.28 μm. Emission around 3.36 μm is characteristic of saturated bonds (aliphatic chains of type

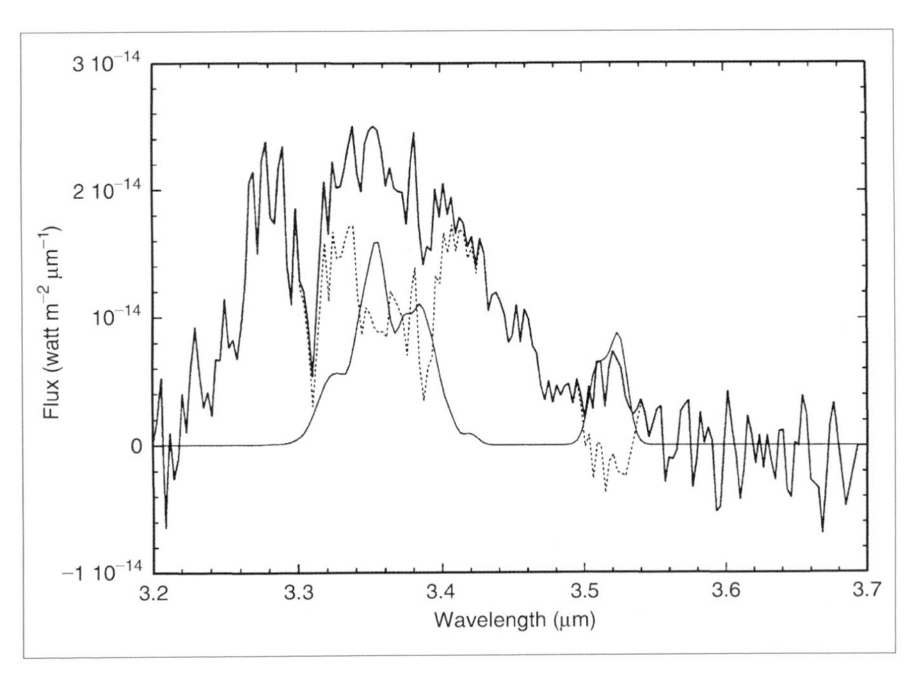

Figure 6.3.
Spectrum in the region around 3 μm for comet Levy 1990 XX, observed using the United Kingdom Infrared Telescope (UKIRT) (*bold curve*). The *thin curve* represents a theoretical spectrum for methanol and the *dotted curve* represents the residual emission after subtraction of the methanol contribution. The residual emission exhibits bands at 3.28 μm (possibly from aromatic compounds), and also at 3.35 μm and at 3.43 μm, whose origins remain unknown. From Davies *et al.*, 1991 [40] and Bockelée-Morvan *et al.*, 1995 [31].

$CH_3—CH_2—CH_3$). Emission at 3.28 μm is characteristic of organic molecules containing unsaturated carbon (that is, some of the carbon atoms have double or triple bonds, as in the alkenes and the alkynes). Such is the case for aromatic compounds, whose molecules contain one or more benzene rings. These are the so-called PAHs, the simplest representatives of which are benzene C_6H_6 and napthalene $C_{10}H_8$. Such molecules have the property that they absorb ultraviolet photons without being destroyed. They convert the photon energy into vibrational energy and then emit infrared photons corresponding to their various vibrational modes. The mechanism has a particularly high efficiency and a very small quantity of aromatic molecules (corresponding to an abundance lying between 10^{-5} and 10^{-4} relative to water) would explain the observed band.

The remainder of the emission residue in Fig. 6.3 has not yet been firmly identified. It does not correspond to any known spectrum of a simple single molecule, and may be due to a mixture of molecules made up of C, H and O, with an abundance of a small percentage.

6.1.4 New molecules observed in comets Hyakutake and Hale–Bopp

Comets Hyakutake and Hale–Bopp were the first really bright comets to benefit from development of a new generation of infrared spectrometers, the cryogenic échelle spectrometers. These instruments have high sensitivity and high spectral resolution and now equip large infrared telescopes such as the United Kingdom Infrared Telescope (UKIRT) and the NASA Infrared Telescope Facility (IRTF), both at Mauna Kea in Hawaii.

One of the major outcomes of these observations was a definite identification of three hydrocarbons: methane CH_4, acetylene C_2H_2 and ethane C_2H_4. They were first identified in comet Hyakutake (Fig 6.4), and the discovery was soon confirmed in comet Hale–Bopp. Detection of ro-vibrational lines of several other species – CO, HCN, OCS, NH_3 – came as no surprise, since these molecules were already known from radio (see below) and UV (for CO) observations. In addition, several lines of as yet undetermined origin have been noted.

Some ro-vibrational lines of water were also observed. These were the first water lines to be observed from the ground. The lines correspond to intermediate steps in the fluorescence cascade which follows excitation to high vibrational levels. Their lower level is not the lowest possible vibrational level, occupied by most water molecules in the Earth's atmosphere. Therefore, these lines are only moderately affected by atmospheric absorption.

6.1.5 ISO results

The apparition of comet Hale–Bopp coincided with operation of the Infrared Space Observatory

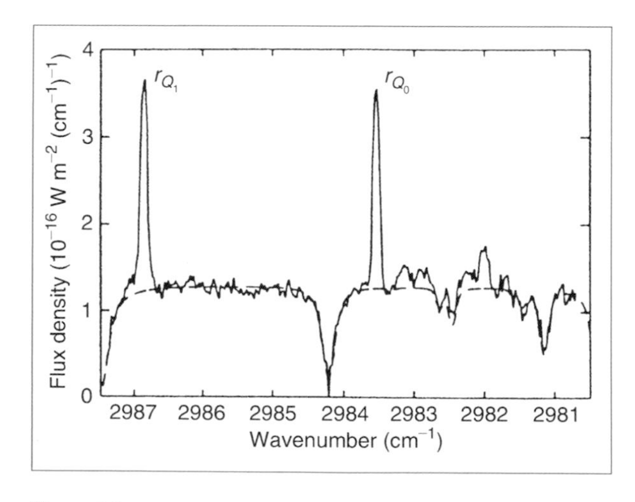

Figure 6.4.
Infrared spectrum of comet C/1996 B2 (Hyakutake) observed with the cryogenic échelle spectrometer of NASA–IRTF, and showing ethane lines. From Mumma *et al.*, 1996 [68].

(ISO). ISO was a follow-up to IRAS, equipped with extensive spectroscopic and imaging capacities (see box entitled *The Infrared Space Observatory*). Unfortunately, there were visibility constraints on ISO and the comet could not be observed while it was within 2.9 AU from the Sun (i.e. at its brightest). The first spectrum of the comet taken by ISO, when it was at 4.6 AU from the Sun, only revealed the 4.25 μm carbon dioxide band (this was the second time this band was observed in a comet, after its detection by Vega-IKS in P/Halley) and the silicate band around 10 μm. Later on, when the comet was at 2.9 AU from the Sun, the whole of its infrared spectrum from 2.4 to 190 μm was observed with good spectral resolution. H_2O, CO and CO_2 bands were revealed, and CO and CO_2 production rates were evaluated at 70% and 22% that of water, respectively. This showed that these molecules are, in addition to water, two important constituents of cometary ice. The water spectrum was studied in detail. Fundamental vibrational bands at 2.7 and 6.5 μm were observed, as well as several rotational lines located in the far infrared. In particular, the 2.7 μm region was observed with great accuracy (Fig. 6.5). By resolving this band into individual ro-vibrational lines, we can study the physical conditions of the water. Its temperature in the coma was thus evaluated at 28 K, which is in agreement with predictions from thermodynamical models of cometary atmospheres (see Chapter 7).

Other important information can be obtained from this spectrum. According to quantum mechanics, water molecules can exist in two states known as *ortho* and *para*, depending on whether the spins of the two hydrogen atoms are parallel or anti-parallel. It is as though there were two water species and that one could only transform into the other by chemical reaction. The distribution of water molecules between these two states, i.e. the ortho to para ratio, depends only on the temperature at which the

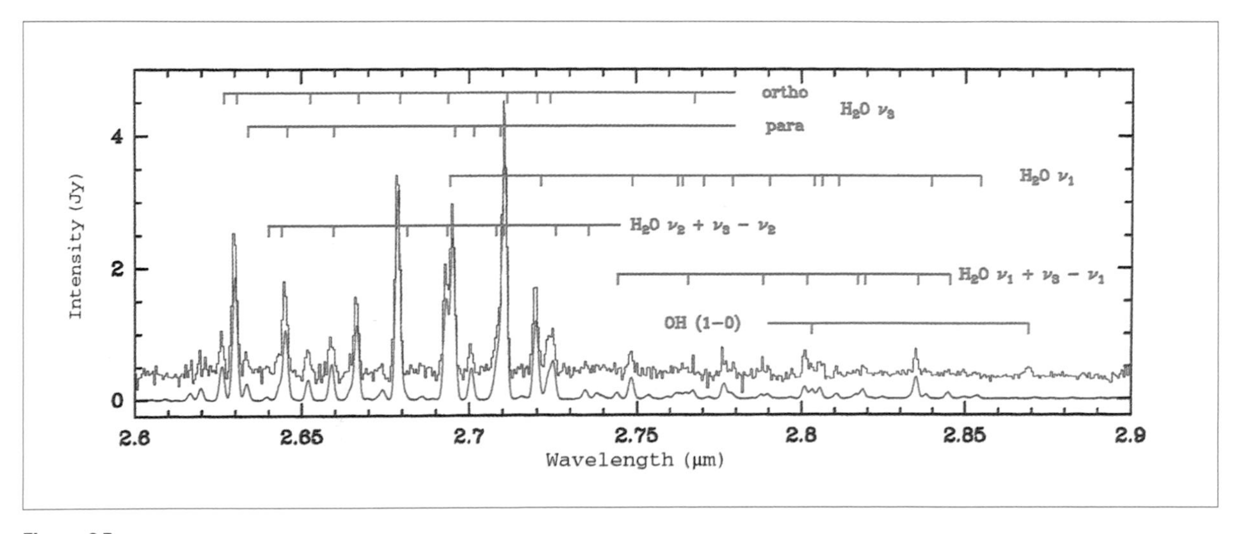

Figure 6.5.
ISO-SWS spectrum of C/1995 O1 (Hale–Bopp) observed on 6 October 1996, showing the H_2O band at high resolution. *Upper*: Observed spectrum. *Lower*: Predicted spectrum. Lines corresponding to ortho and para water states are marked. Their relative intensities correspond to a formation temperature of around 25 K. From Crovisier *et al.*, 1997 [39].

molecules formed and is unchanged over time. Water molecules thus keep a souvenir from the time they were formed. In the Hale–Bopp H_2O spectrum (Fig. 6.5), both ortho and para water lines are visible. We deduce that the cometary water was formed at a temperature of about 25 K. Such a low temperature suggests that formation occurred either in interstellar clouds or in some outer region of the protoplanetary solar nebula.

Between 7 and 45 μm, the Hale–Bopp spectrum (Fig. 6.6) corresponds to a blackbody upon which are superposed a certain number of bumps. Thermal emission from cometary dust produces the blackbody spectrum, whose shape indicates a temperature of about 200 K. This is more or less the temperature expected at that heliocentric distance (see Fig. 6.1). Bumps characterise the composition of the cometary dust.

Emission around 10 μm is a silicate feature, and has been known for a long time in cometary dust, as well as in many other sorts of cosmic dust. The silicates form a wide class of minerals, many of which are found in terrestrial rocks. Amorphous (glassy) silicates produce only two wide bands towards 10 and 20 μm, whereas crystalline silicates have several other bands whose positions depend on their composition. Further bands observed in the Hale–Bopp spectrum at 16, 23.5, 27.5 and 33.5 μm are very close to the signature of magnesium-rich crystalline olivine, with formula Mg_2SiO_4. This mineral, called forsterite, is well known on Earth.

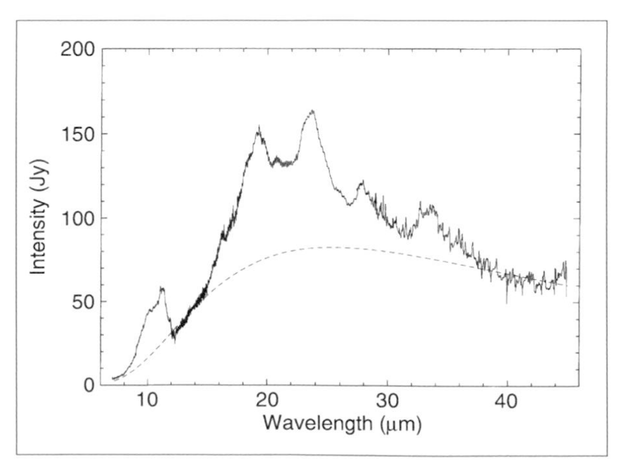

Figure 6.6.
The 7–45 μm ISO-SWS spectrum of C/1995 O1 (Hale–Bopp) observed on 6 October 1996 is essentially due to dust emissions. The *dotted line* shows the 200 K blackbody spectrum for comparison. Bumps are spectral signatures of forsterite, a magnesium-rich crystalline silicate which is one of the components of cometary dust. From Crovisier *et al.*, 1997 [39].

6.2 Radio observations

6.2.1 Historical beginnings: the OH radical

Radio astronomy is a method of observation which came to the fore in the 1950s, inheriting from radar techniques developed during the Second World War. After several unsuccessful applications to 'historic' comets (Arend–Roland 1957 III, Ikeya–Seki 1965 VIII, Bennett 1970 II), the first positive results were finally obtained for comet Kohoutek (1973 XII). The apparition of this comet, which was bright without being spectacular, coincided with the setting up of the manned American orbital space station, Skylab. For this occasion, NASA had encouraged a large international observation campaign, which was the first of its kind, and which foreshadowed the Halley campaign.

It was for this comet that the 18 cm lines of the OH radical were first recorded using the Nançay radio telescope (France). The result was confirmed by the 42 m radio telescope at Green Bank (USA). As this radical is produced when water molecules decompose under the influence of solar UV radiation, these observations give an almost direct indication of a comet's global gas

The Infrared Space Observatory (ISO)

The Infrared Space Observatory was launched on 17 November 1995 by the European Space Agency. Its telescope had a 60 cm mirror and was contained in a liquid helium cryostat (see Fig. 6.7). ISO was in operation until 8 April 1998, at which point all the helium had completely evaporated. ISO was constrained to point its telescope between 60 and 120° from the Sun, to maintain an acceptable insulation of its solar panels and protect its thermal equilibrium. The Moon and Earth had also to be avoided. This seriously limited the region of sky accessible to ISO observations at any given time, and complicated observation planning. Certain objects in the Solar System could not be viewed, such as Mercury, Venus, and comets passing close to the Sun.
Several instruments were available in the ISO focal plane:

- **SWS** (Short Wavelength Spectrometer): Operating between 2.4 and 45 μm, with resolving power about 1500 (grating) and 30 000 (Fabry–Pérot).
- **LWS** (Long Wavelength Spectrometer): Operating between 45 and 190 μm, with resolving power about 200 (grating) and 10 000 (Fabry–Pérot).
- **ISOCAM** (ISO camera): A camera equipped with a detector array, a filter set and a circular variable filter (resolving power 50). This was capable of obtaining images in different infrared 'colours' and even low resolution spectra.
- **ISOPHOT** (ISO photometer): A multi-purpose photometer, capable of accurate and sensitive flux measurements in various colours and with various diaphragms. It could also measure low resolution spectra between 2.5 and 12 μm (resolving power 90).

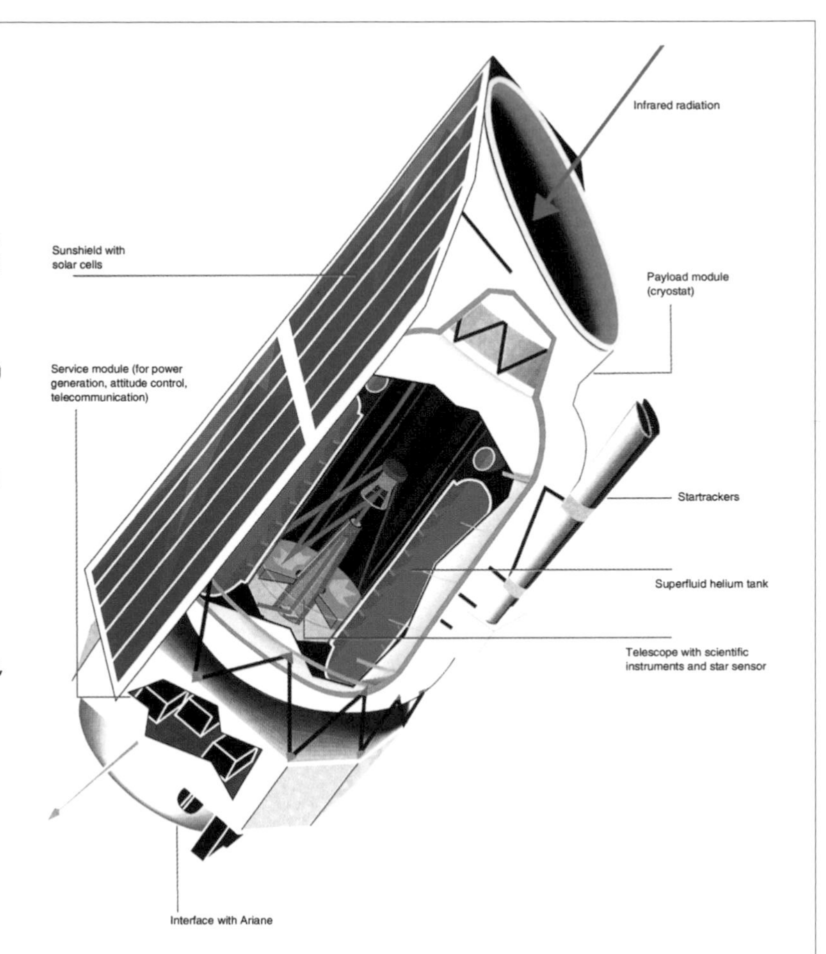

Figure 6.7.
The Infrared Space Observatory (ISO).

production. Systematic observations of the OH 18 cm lines were subsequently undertaken at the Nançay radio telescope for several dozen comets. Radio observations at these wavelengths can be made equally well during the day as during the night, and without regard for prevailing weather conditions, which represents a considerable advantage over optical observations.

Other attempts to make radio observations produced less convincing results at this period. It was not until the beginning of the 1970s that centimetre and millimetre radio astronomy made its first great breakthroughs in the search for interstellar molecules, when telescopes equipped with efficient millimetre detectors were put into operation. Radio astronomers hoped to carry out

Figure 6.8.
300 × 35 m spherical mirror of the large Nançay radio telescope (Cher, France). It has observed the radical OH 18 cm lines in almost fifty comets. Photo by J. Bérezné, courtesy of CNRS.

for comets those observations which had been fruitful for the interstellar medium. However, several detections announced for comet Kohoutek (HCN, CH_3CN, and the CH radical) could not be repeated for this comet, and nor were they confirmed for the bright comets of subsequent years.

6.2.2 Parent molecules

Following these promising results, no other positive result was obtained, except for OH observations, before the return of comet Halley. At the time of Halley's return, a 30 m radio telescope had just been built at altitude 2800 m in the Spanish Sierra Nevada. This was a project of the Franco-German research institute IRAM (Institut de radioastronomie millimétrique), which Spain was later to join. It was the biggest telescope operating at millimetre wavelengths. In November 1985, it detected hydrogen cyanide (HCN) lines, during the first few observations of comet Halley (see Fig. 5.7). The discovery was soon confirmed by American and Swedish teams. These lines at wavelength 3 mm had been suspected, although less convincingly, in comet Kohoutek. They finally identified a parent molecule responsible for the CN radical, so clearly witnessed in visible cometary spectra. Using the Very Large Array (VLA, a network of 27 dishes, each of 25 m diameter, in New Mexico), a centimetre line of formaldehyde (H_2CO) was also detected in P/Halley, although marginally.

Lacking any suitable target objects, no further discoveries were made until four years later, when comets Austin 1990 V and Levy 1990 XX arrived on the scene. Formaldehyde was identified by its millimetre lines, together with two

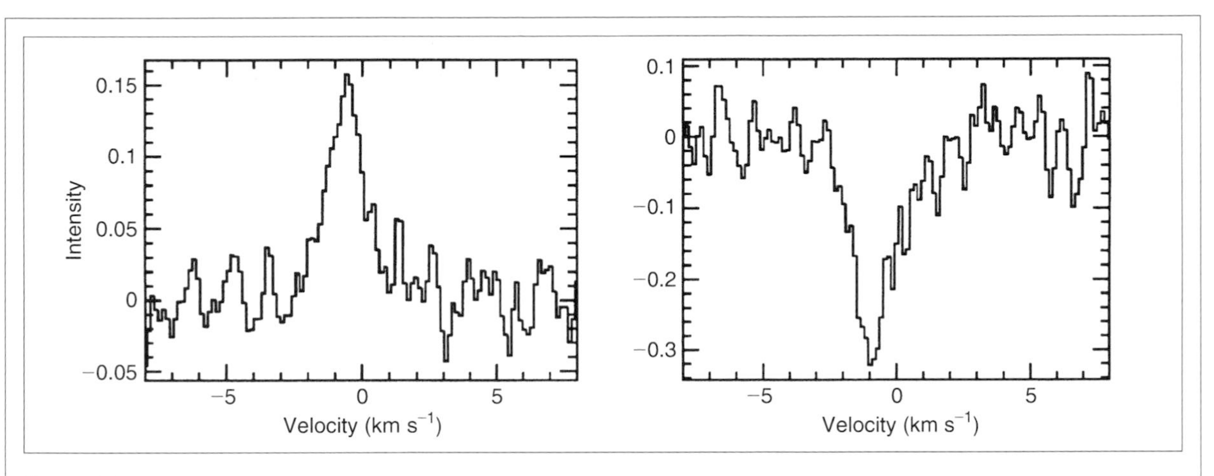

Figure 6.9.
18 cm OH lines observed in the spectrum of comet P/Swift–Tuttle 1992 XXVIII. On 29 October 1992 (*left*), an emission line was observed, whereas on 1 December 1992 (*right*), an absorption line appeared. This change was due to a difference in excitation, which was in turn caused by variation in the comet's heliocentric speed: –17 km/s on 29 October and –5 km/s on 1 December. This is known as the *Swings effect*. From the cometary spectrum collection at Nançay radio telescope (France).

The 18 cm OH lines

The 18 cm lines are not rotational lines. For a light molecule such as OH, rotational lines lie in the submillimetre and far infrared regions (see Chapter 4). As a result of the electronic structure in the unfilled shells of this molecular radical, each energy level is split: this is referred to as *lambda splitting*. The 18 cm lines are produced by transitions between these two sublevels, for the lowest energy state (electronic, vibrational and rotational) of the OH molecule. The molecules are usually distributed evenly between sublevels, so the 18 cm lines are rather faint, and undetectable in comets. However, these molecules are *pumped* by solar radiation, through the fluorescence mechanism which gives rise to the strong emission band in the near UV (see Chapter 4). Such pumping leads to overpopulation of either the lower or higher sublevel, depending on the comet's heliocentric speed. This is known as the *Swings effect*, already encountered in Chapter 4. The OH molecules are no longer in equilibrium and constitute a *maser* (Microwave Amplification by Stimulated Emission of Radiation). If the upper sublevel is overpopulated, the sky background 18 cm radiation (the 3 K cosmic radiation, to which the galactic continuum radiation is sometimes added) is amplified, resulting in an *emission* line. If, on the other hand, the lower sublevel is overpopulated, this radiation is attenuated and an *absorption* line is observed. Cometary OH lines are indeed observed either in emission or in absorption (see Fig. 6.9), depending on the comet's velocity, in agreement with this model.

new parent molecules. Firstly, a sulphur-containing molecule was detected, namely hydrogen sulphide (H_2S). Sulphur had only previously been observed through CS, S_2 and ultraviolet atomic sulphur lines. Secondly, methanol (CH_3OH) was found, with abundance ratio greater than 5% relative to water in certain comets. It thus appears to be one of the most abundant volatile components of cometary nuclei. These observations were repeated on comet P/Swift–Tuttle in 1992, with even greater precision, thanks to the intensity of the lines. The wavelength range was extended down to the submillimetre region where further HCN, H_2CO and CH_3OH lines were observed using the JCMT (James Clerk Maxwell Telescope). This is a 15 m submillimetre telescope located at the summit of Mauna Kea, Hawaii.

During the winter of 1993–94, a radio line of carbon monoxide (CO) was detected in the un-

Figure 6.10.
30 m IRAM radio telescope on Pico Veleta, in the Sierra Nevada, Spain (altitude 2800 m). This telescope is actively engaged in the hunt for interstellar and cometary molecules. Photo by J. Crovisier.

usual comet P/Schwassmann–Wachmann 1, once again with the JCMT. This comet follows an almost circular orbit at 6 AU from the Sun, beyond the orbit of Jupiter. It is well known for its bursts of activity, as sudden as they are unexpected, during which its brightness increases by several magnitudes. It thus belongs to a group of comets showing activity at large heliocentric distances, e.g. P/Halley at 14 AU, Chiron at 10 AU, and several others. Such activity cannot be explained by water sublimation. Carbon monoxide would seem to be responsible for this activity, although there are other candidates, like N_2 and CO_2. Radio astronomy is particularly well suited to this type of study. It can detect rotational lines of molecules in cold media, whereas vibrational and electronic bands must be excited by solar radiation before they can emit by fluorescence. The latter would be too faint to detect at such large heliocentric distances, even with highly sensitive equipment, like the Hubble Space Telescope.

Comets Hyakutake and Hale–Bopp were particularly rewarding in the search for new molecules at radio wavelengths. Almost all the major radio telescopes took part in these observations (Fig. 6.11).

Among newly discovered organic molecules are isocyanic acid (NH_2CHO), acetaldehyde (CH_3CHO) and methyl formate ($HCOOCH_3$) (perhaps the largest molecule firmly identified in a comet to date). Suspected for a long time to be a major cometary volatile, ammonia (NH_3) has at last been firmly identified by its rotational lines at wavelength 1.3 cm (it is the only cometary molecule observed in this spectral range). Nitrogen is further represented by several nitriles: in addition to hydrogen cyanide (HCN), its isomer (HNC) was found, as well as methyl cyanide (CH_3CN) and cyanoacetylene (HC_3N).

Several sulphur-bearing molecules are now known. H_2S, CS and S_2 had already been observed: CS, presumably a decomposition product of CS_2, had been observed in the UV, and S_2 had been observed in the near UV, although only in IRAS–Araki–Alcock 1983 VII and C/1996 B2 Hyakutake when these two comets made very close approaches to Earth. In addition to these molecules, sulphur monoxide (SO), sulphur dioxide (SO_2), carbonyl sulphide (OCS) and thioformaldehyde (H_2CS), the sulphur-bearing analogue of formaldehyde, were also revealed.

Apart from these parent molecules, cometary ions were observed, for the first time at radio wavelengths: CO^+ (already known from its lines

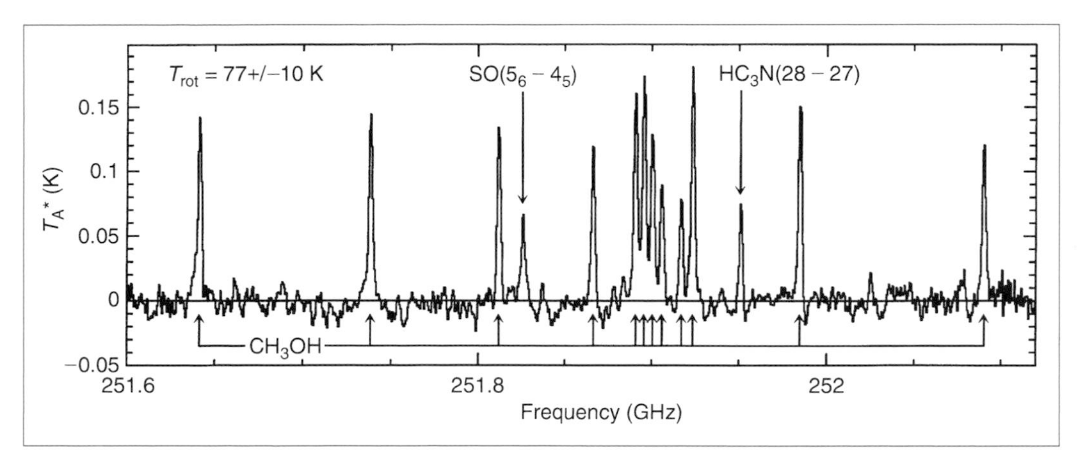

Figure 6.11.
A Hale–Bopp spectrum observed with the Caltech Submillimeter Telescope in February 1997, showing SO and HC_3N lines, and a series of methanol (CH_3OH) lines. From Lis *et al.*, 1998 [62].

in the visible spectrum of cometary tails), HCO^+ and H_3O^+. These ions allow us to study the chemistry of cometary atmospheres, dominated by photolysis and ion-neutral reactions. Their line shapes are very different from those of cometary neutral species, because ions are accelerated in solar wind interactions.

Finally, several isotopic species were also observed. Deuterated water (HDO) was detected through one of its submillimetric lines at the Caltech Submillimeter Observatory (CSO) in Hyakutake (Fig. 6.12) and with the JCMT in Hale–Bopp. Deuterated hydrogen cyanide (DCN) and isotopic species of carbon and nitrogen (in HCN) and of sulphur (in CS) were also observed. These results will be further discussed in Chapter 9.

Much is expected from the gradual extension of radio astronomy towards submillimetre wavelengths. The fundamental (and hence the most intense) rotational lines of light molecules, such as hydrides (e.g. H_2O, NH_3), lie in this region. Water is a case in point. Its fundamental line, between the two lowest rotational energy levels, lies at frequency 557 GHz (or wavelength 0.5 mm). It is inaccessible to Earth-based observation. The Earth's atmosphere is opaque to this frequency, precisely because it contains water itself. This line is so important in astrophysics that small space radio telescopes are at present being studied, specifically for its observation (see Chapter 11).

Radio spectroscopy has thus proved to be a very efficient tool in identifying cometary parent molecules. Because it can observe the frequency of narrow lines with very great accuracy, it leads to unambiguous identification of molecules (other techniques at other wavelengths sometimes lead to indecipherable forests of lines in which signals from several molecular species blend together). It is also very sensitive, since species with mixing ratios as low as 10^{-4} relative to water, or even less, have been observed in the two recent comets. Radio spectroscopy, however, is blind to non-polar molecules, such as CO_2 or CH_4, which do not have allowed rotational transitions.

When complemented by infrared observations (described above), radio observations of comets Hyakutake and Hale–Bopp now provide us with

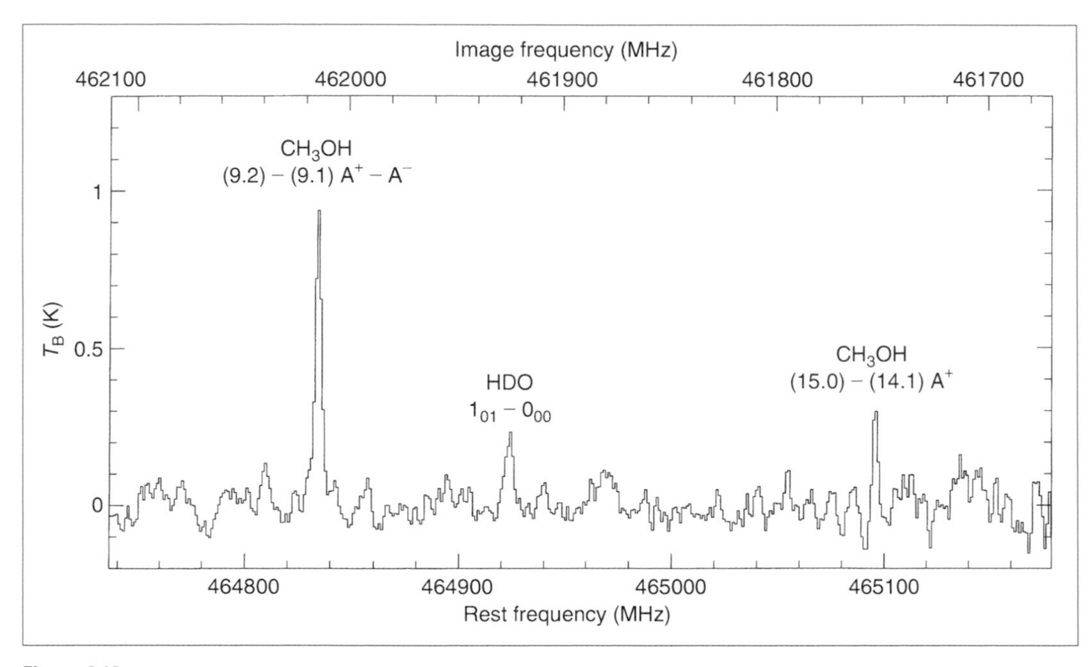

Figure 6.12.
Deuterated water (HDO) line observed in comet Hyakutake with the Caltech Submillimeter Telescope. Two methanol lines are also apparent in the same spectrum. From Bockelée-Morvan *et al.*, 1998 [32].

a fairly comprehensive inventory regarding the composition of cometary volatiles. This will be discussed further in Chapter 9.

6.2.3 Radio line profiles and kinetics of cometary atmospheres

One of the advantages of radio observations is their high spectral resolution. Line profiles reflect the velocity distribution of molecules emitting them: because of the Doppler effect, the frequency of a spectral line is shifted by $\Delta\nu$ relative to its rest frequency ν, where $\Delta\nu/\nu = -v/c$, v is the velocity of the molecule relative to the observer and c is the speed of light. In cometary atmospheres, molecular velocities are the resultant of atmospheric expansion velocities (around 1 km/s) and velocities of thermal motion within the gas (a fraction of 1 km/s). This leads to relative shifts of $v/c \approx 1/300\,000$. Visible, UV and infrared spectroscopy techniques are quite incapable of observing such effects. At radio wavelengths, however, spectral resolutions are practically limited only by noise in the signal and receiver themselves. Using coherently connected receivers (heterodyne), fitted with filters or autocorrelation spectrometers (based on Fourier transforms), line profiles can be measured with great accuracy (see Fig. 6.13). Of course, line profiles include speeds of all molecules contained within the radio telescope field (a fraction of an arcmin). This corresponds to tens of thousands of kilometres for a comet at 1 AU. The expansion speed of the atmosphere can nevertheless be deduced by comparing observed profiles with model predictions.

In many cases, line profiles are neither symmetric, nor centred on the speed of the cometary nucleus (see Fig. 6.13). This is due to bulk motion of cometary gases relative to the

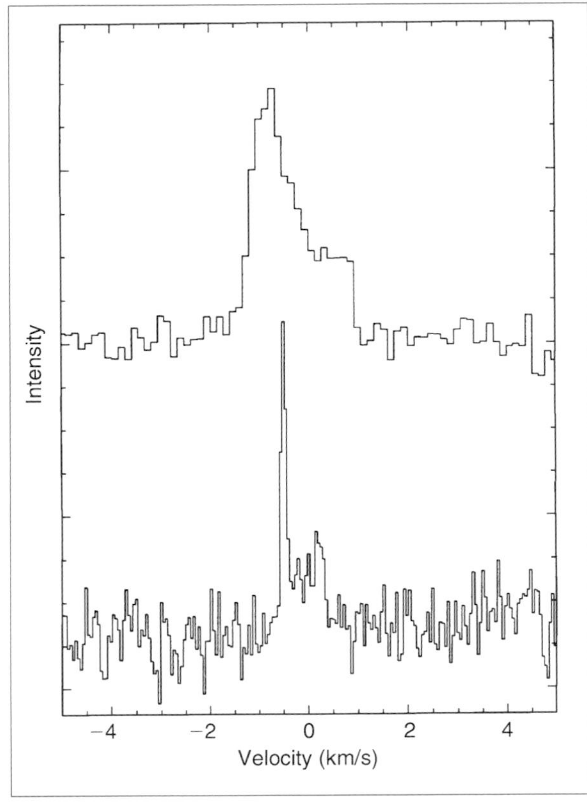

Figure 6.13.
Molecular line profiles observed with the IRAM 30 m radio telescope. *Upper*: Formaldehyde (H_2CO) line in comet P/Swift–Tuttle at 1 AU from the Sun. *Lower*: Carbon monoxide (CO) line in comet P/Schwassmann–Wachmann 1 at 6 AU from the Sun. Note that the line is much wider in the first comet. The activity of P/Swift–Tuttle is dominated by water sublimation, while that of P/Schwassmann–Wachmann 1 is probably dominated by CO sublimation, at a much lower temperature. It follows that the expansion speed of the coma, and hence the line width, is much greater in the first case. For both comets, an intensity peak is observed at negative velocities. This implies a preferential gas emission towards the observer, i.e. from the nuclear face heated by the Sun. From Crovisier *et al.*, 1995 [38] and Despois *et al.*, 1995 [43].

nucleus. These gases are not escaping by a spherical radial expansion, but rather in jets, ejected primarily from the hot face of the nucleus, exposed to the Sun. As we shall see in Chapter 7, the reaction force exerted by these jets on the nucleus can explain non-gravitational effects in cometary motion. Radio observations which reveal jets and measure their velocities are therefore useful in quantifying such effects.

6.2.4 Radio interferometric observations

An efficient way of making images at radio wavelengths is to use radio interferometers, in which signals from several (small) radio telescopes are combined. The resulting map has a resolution which can be as small as λ/B, where λ is the wavelength and B the base line, or distance between individual telescopes. Current millimetre radio interferometers, with baselines of about half a kilometre, have resolutions of 1–3 arcsec. For comparison, the resolution of a single dish antenna is λ/D, where D is now antenna diameter, and this is typically 10–30 arcsec for current millimetre radio telescopes. This technique involves making observations at different times and varying distances between telescopes, and is therefore not so easy to apply to variable objects like comets.

Radio interferometry was first used to map the distribution of the OH radical at 18 cm in comet Halley, using the 27 antennas of the Very Large Array (VLA). Recently, the IRAM interferometer at Plateau de Bure in the French Alps (5 antennas of 15 m diameter operating at millimetric wavelengths) was able to map HCN and CO distributions in comet Hyakutake, as well as distributions of half a dozen species in comet Hale–Bopp (see Fig. 6.14), with a spatial resolution of one or two arcsec. Their origin and distribution in the coma could then be traced.

Radio interferometers have also proved to be highly efficient in detecting new molecules by simply adding the signals from their multiple antennas. This method applied at the Plateau de Bure interferometer led to the discovery of CH_3CN in comet Hyakutake, and also SO_2 and formic acid in comet Hale–Bopp. The Berkeley–Illinois Millimeter Array (BIMA) likewise

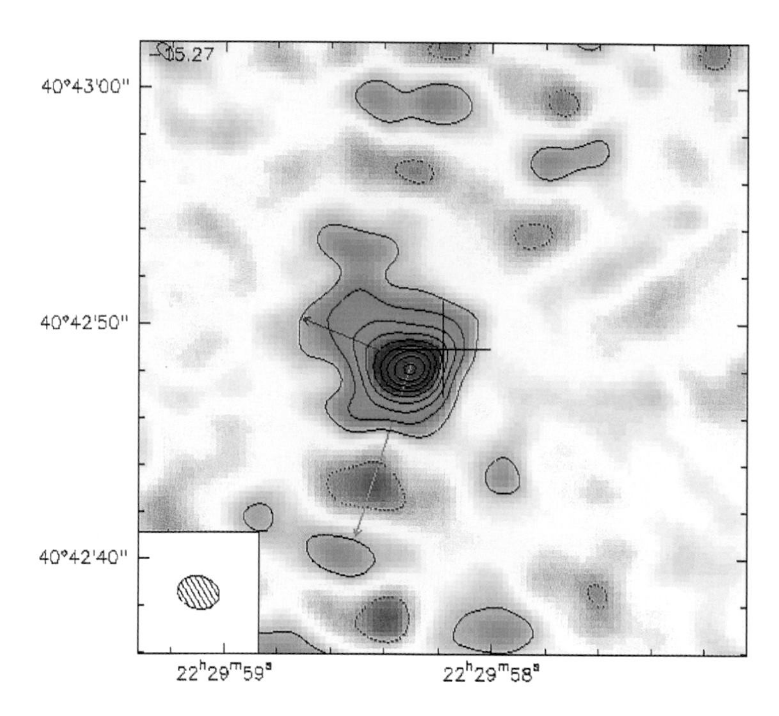

Figure 6.14.
The map of the CO J(2-1) line at 230 GHz, observed in comet Hale–Bopp on 11 March 1997 with the IRAM interferometer at Plateau de Bure (France). The ellipse at the bottom-left corner shows the spatial resolution of the instrument. The blue arrow shows the direction of the Sun, and the black arrow that of the comet motion. Courtesy of IRAM.

obtained the first detection of the HCO^+ cometary ion.

6.2.5 Continuum and radar observations

Continuum radio emission from comets is due to thermal emission from their nucleus and surrounding dust. It has only been detected with certainty in a limited number of cases, including comet Halley and a few of the comets appearing since. Detections were made at millimetre wavelengths (1.3 mm) using the IRAM radio telescope, and at submillimetre wavelengths (0.8 mm) using the JCMT. Further detections, at centimetre wavelengths have been announced, but it has not been possible to confirm them.

Continuum radiation from cometary nuclei is very faint, because they are so small. It can only be observed for the larger comets, and even then only those passing near to Earth. The same is true for asteroids; only those measuring several hundred kilometres in diameter, or passing exceptionally close to Earth, have been observed in this way. The radiation observed from comet Halley was clearly greater than could be explained by its nucleus alone. Excess emission may be due to cometary dust. However, small particles cannot radiate efficiently at wavelengths greater than their own dimensions. Emitting grains are not, therefore, 'visible' cometary dusts, whose size must be less than one micron. They are doubtless grains of several millimetres, or even larger 'pebbles', which surround the nucleus. A halo of ice grains has also been invoked to explain this emission. But at a heliocentric distance of 1 AU, such grains would melt like snow in the Sun, and calculations show that they would have an extremely short lifetime.

Radar techniques have been useful in studying certain objects in the Solar System, but are severely limited by distance. The return signal is proportional to S/Δ^4, where S is the cross-sectional area of the object and Δ its distance. The largest possible antennas must be used, such as the 70 m antenna at Goldstone, California, or the 300 m radio telescope at Arecibo, Puerto Rico. Both instruments are equipped with powerful centimetre and decimetre emitters. Apart from planets and their larger satellites, several dozen asteroids have been observed, either in the main asteroid belt (only the larger ones) or among those passing very close to Earth (e.g. Toutatis, in 1992). Likewise for comets, we are limited to those with large nuclei passing very close to Earth. For this reason, only a small number of comets have been investigated by radar. Echoes have been obtained from P/Encke and P/Grigg–Skjellerup, at 0.33 AU from Earth, from IRAS–Araki–Alcock (although small, it passed at only 0.031 AU from Earth in May 1983), from P/Halley at 0.63 AU, and from Hyakutake at 0.10 AU from Earth (see Fig. 6.15).

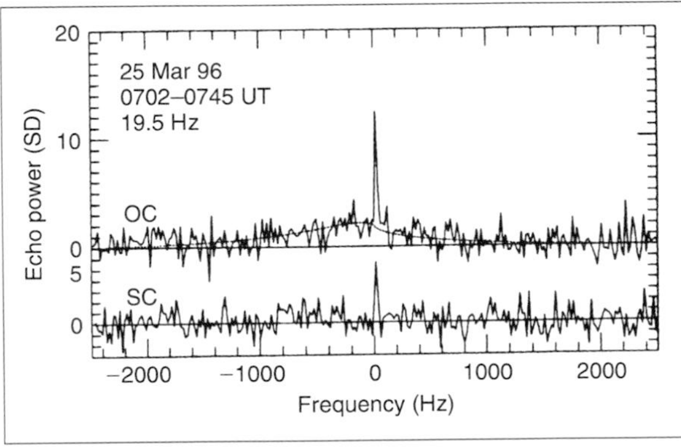

Figure 6.15.
The radar echo obtained at the Goldstone antenna from comet Hyakutake at the moment of its closest approach to Earth. The sharp spike is due to reflection from the nucleus, for which a diameter of 2 to 3 kilometres was derived. The broader component is due to backscatter by large grains ejected from the nucleus at speeds of several tens of metres per second. The *abscissa* represents the doppler shift of the signal (relative to that expected from the velocity of the comet nucleus). SC denotes the same direction of circular polarisation as that of the outgoing radar signal, and OC the opposite direction. For reflection on a smooth solid surface, the direction of polarisation is reversed. The presence of an SC signal indicates reflection on a rough or porous surface. From Harmon *et al.*, 1997 [50].

Radar observations give precise distance measurements, by echo delays, and also determine velocities, through Doppler shifting of echo frequencies. This is useful in accurately ascertaining orbits. They also help to establish limits on size, reflection properties and rotation of objects under study, through echo intensity, shape and polarisation. Although they cannot alone completely determine all these parameters, it is sometimes possible to combine information from other sources. In the case of comets IRAS–Araki–Alcock, P/Halley and Hyakutake, observed signals were incompatible with predictions based on a compact nucleus. They seem to be the result of echoes from an ensemble of large grains (of dimensions around one centimetre) surrounding the nucleus. This would tend to confirm the idea that continuum radio emission from comets is due to a cloud of such large grains.

6.3 X-ray observations

A totally unexpected result from the comet Hyakutake observing campaign was the discovery that comets could be important X-ray sources. This discovery was soon confirmed in comet Hale–Bopp, and in several other comets from archive data. This phenomenon is therefore not rare and could be observed even in some faint comets. So far, cometary X-ray emission has been observed by no fewer than five X-ray satellite observatories, initially built to observe high energy emissions from the Sun or galactic and extragalactic sources:

* The *Röntgen X-ray Satellite* (ROSAT) (0.1–2.4 keV) was the first to discover X-rays in comet Hyakutake [63]. Comet Hale–Bopp was subsequently observed and several other comets were retrieved from archive data.

* The *Rossi X-ray Timing Explorer* (XTE) also participated in the discovery of X-rays in comet Hyakutake [63].
* The *Extreme Ultraviolet Explorer* satellite (EUVE) (soft X-rays: 70–180 Å, or 180–70 eV) also observed comets Hyakutake and Hale–Bopp as well as several other comets [58].
* *ASCA* and *BeppoSAX* observed comet Hale–Bopp [72].

X-rays are high-energy radiation which cannot be emitted by an inert, cold body such as a comet on its own. This emission can only come from interactions of cometary matter with high energy solar radiation or solar particles. The exact nature of this interaction, however, has not yet been assessed. It could be charge transfer from heavy ions in the solar wind to neutral atoms in the cometary atmosphere; cometary atoms would be excited to high electronic levels and their de-excitation would result in X-ray emission. Or it could be scattering of solar X-ray emission by very small cometary grains (attogram particles, i.e. dust grains of 10^{-18} g). Other mechanisms have also been proposed.

7
Cometary nuclei and their activity

Preceding chapters have dealt with methods, both direct and indirect, for obtaining information about comets and understanding their origins and evolution. With these tools, we can now approach the fundamental questions in cometary physics: What is a cometary nucleus? What is it made of? How did it form and how does it evolve? What is its relation to other small bodies in the Solar System? And finally, what can it teach us about the origins of the Solar System?

7.1 Our knowledge of cometary nuclei

There are only a very few comets for which any features of the nucleus are known, even approximately. These are summarised in Table 7.1, together with properties of some other small bodies in the Solar System, to be discussed in Chapter 10. With the exception of comet Halley, whose nucleus has actually been photographed by space probes, the radius, albedo and rotation period are deduced from photometric observations of cometary light curves. These observations are made when the comet is far from the Sun, so that it is practically inactive (see Chapter 3). It is worth recalling that quite often the size and albedo of the nucleus cannot be determined independently. In a very small number of cases, it has been possible to deduce the nucleus size from radar observations (see Chapter 6). In several other cases, a rotation period has been deduced from observation of dust jets. These evolve rather like water jets from a rotary sprinkler (e.g. for comet 109P/Swift–Tuttle, shown in Fig. 3.9, and recently in comet Hale–Bopp). The *active fraction* of the nucleus is defined as the proportion of its surface area covered with exposed ice. It is related to the water production rate observed when a comet is active.

It is not easy to determine the densities of cometary nuclei. This requires knowing their size and their mass. It will only be possible to measure the mass of such an object directly when we can place a space probe in orbit around one, as proposed in the Rosetta project (see Chapter 11). However, it is sometimes possible to estimate the mass indirectly by modelling non-gravitational forces. Indeed, we can calculate perturbations to the comet's orbit when sublimating gases are ejected anisotropically from the nucleus. Unfortunately, many poorly determined parameters come into play and uncertainty in density estimations leaves us with a range of values from 0.1 to 1 g/cm^3 (see the box entitled *Non-gravitational forces and densities of cometary nuclei*).

We have seen how a gathering body of information led to the hypothesis, formulated in the 1950s, that activity in cometary nuclei is domi-

Table 7.1. Physical properties of cometary nuclei and other small bodies[a] in the Solar System.

Object	Radius[b] (km)	Rotation period (h)	Albedo	Active fraction
Comets[c]				
C/1995 01 (Hale–Bopp)	20–40	11.3		
29P/Schwassmann–Wachmann 1	20	complex?	0.13	0.06
109P/Swift–Tuttle	12	72.		0.03
28P/Neujmin 1	10	12.7	0.03	0.001
1P/Halley	8 × 4	complex	0.04	0.20
10P/Tempel 2	8 × 4	8.9	0.02	
IRAS–Araki–Alcock 1983 VII	8 × 3			0.006
49P/Arend–Rigaux	7 × 4	13.6	0.03	0.001
2P/Encke	~ 2	15.1		
C/1996 B2 (Hyakutake)	1–1.5	6.3		~ 1
46P/ Wirtanen	~ 0.6	6?		~ 1
Asteroids				
(1) Ceres	456	9.1	0.10	
(243) Ida	28	4.5		
(4179) Toutatis	2.3	complex		
Centaurs				
(2060) Chiron: 95P/Chiron	90	5.9	0.13	
(5145) Pholus	~ 100			
Trans-Neptunian objects				
Pluto	1150	153	0.61	
1992 QB_1	~ 100			

[a]Listed here for comparison. To be discussed in Chapter 10.
[b]20–40 indicates a range of possible values. 8 × 4 indicates extreme dimensions of an elongated nucleus.
[c]In decreasing order of size.

nated by sublimation of water ice. This is outlined in the box entitled *Sublimation of cometary ices*. The model was confirmed during space exploration of comet Halley, and also by recent spectroscopic observations. It now seems valid at least for comets penetrating the inner regions of the Solar System. However, it would not appear to apply to comets which are active at great distances from the Sun.

Cometary ices are not made of pure water, but rather a mixture of more or less volatile molecules. When heated, this mixture undergoes *fractional sublimation*, i.e. the most volatile molecules are the first to go. A differentiation process occurs in the outer layer of the cometary nucleus, which is depleted in molecules such as CO, CO_2, and N_2. Numerical calculations show that this outgassing phenomenon can be very complex. Proportions of CO, CO_2 and H_2O released by the nucleus depend not only on its distance from the Sun, but also on the comet's past history. Behaviour may be radically different before and after perihelion. A new comet will release its most volatile compounds in a single blast before perihelion, whereas a comet that has already returned several times will only release

Non-gravitational forces and densities of cometary nuclei

If matter, in the form of gas and dust, is not escaping isotropically from the cometary nucleus, it follows by conservation of momentum that there will be a net reaction force on the nucleus. This is given by

$$F = -Qv,$$

where Q is the matter production rate, as a mass per unit time, and v the resultant velocity of this matter as a whole. The latter would be zero if ejection were isotropic.

F is the non-gravitational force which must be added to attractive gravitational forces from the Sun and planets. It can produce an observable effect on the cometary orbit. The most noticeable effect is an advance or a delay in the perihelion passage. For example, P/Halley was 4 days late and P/Brorsen–Metcalf 15 days ahead of schedule, due to this effect. If we can estimate F and observe the non-gravitational orbital effects it produces throughout the orbit, we can then estimate the comet's mass. If in addition the size of the nucleus is known from other observations, its density can be calculated.

The resultant velocity v is estimated to be of the order of 0.2 km/s (something like a quarter of the coma expansion speed). This comes from a combination of theoretical deduction and measurements of Doppler shifts in molecular radio lines. Roughly speaking, it is directed towards the Sun, although it may stray significantly from this direction. Maximum sublimation occurs in the 'afternoon' for a rapidly rotating nucleus, because of thermal inertia. Matter production rates are estimated on the basis of photometric and spectrophotometric measurements of gas and dust production rates. It is particularly important to measure how this production evolves in time. Changes in orbital period are very sensitive to any asymmetry in activity before and after perihelion passage.

Non-gravitational effects are known for orbits of about eighty comets. However, mass estimates have only been possible for a very few objects (e.g. P/Halley, P/Kopff), and estimates by different groups often disagree. P/Halley is the only comet whose nuclear dimensions are reliably known. The estimated density has mean value 0.3 g/cm^3, but estimates spread over a range 0.1–0.9 g/cm^3 owing to great uncertainty in the value and direction of v, and in the value of Q and its evolution.

them after. This is because, in the latter case, volatile components are absent from surface layers, and the Sun's heat must penetrate the nucleus to a certain depth.

Some comets are active at very great heliocentric distances, and this implies the presence of highly volatile compounds. Examples are Chiron which, at 10 AU, has a discernable coma, and P/Halley which produced a burst of activity at 14.7 AU (see Fig. 7.2). The record seems to be held by comet Černis 1983 XII in which activity was recently detected at 23 AU from the Sun. Such phenomena are probably not uncommon, but they are difficult to observe, requiring the use of powerful telescopes. As we saw in Chapter 6, carbon monoxide was detected in comet P/Schwassmann–Wachmann 1 at 6 AU from the Sun, and this may well be the molecule responsible for distant cometary activity.

The early discovery of a very bright comet far from the Sun, comet Hale–Bopp, gave an unprecedented opportunity to study the long term evolution of outgassing from a cometary nucleus. The production of several species was monitored using radio observations, almost from the moment of discovery at 7 AU from the Sun until perihelion at 0.9 AU, and afterwards while the comet was receding from the Sun (Fig. 7.3). Carbon monoxide and water evolved quite differently. CO increased slowly between 6.7 and 3 AU, then stalled before increasing again by 2 AU. H_2O increased very steeply beyond 3 AU, surpassing CO around 3.5 AU. Between 2 AU and perihelion, CO and H_2O showed a similar evolution. We thus witnessed the change between CO-driven and H_2O-driven sublimation regimes. After perihelion, the same scheme was repeated in reverse order. Other species observed followed their own variation patterns, which resembled either CO or H_2O, but lacked any strict correlation with their volatility.

Sublimation of cometary ices

Let us consider the thermal balance of an icy body (water ice, in this case), when it is heated by the Sun. The aim is to calculate its equilibrium temperature T and the rate of sublimation of ice. Making several approximations, we find

$$\frac{F(1-A)}{r_h^2} = \eta(\sigma T^4 + LZ) .$$

The left-hand side is the energy available per unit surface area. F is the solar constant (1360 $\mathrm{Wm^{-2}}$ at 1 AU from the Sun), A the albedo and r_h the heliocentric distance in AU.

The first term on the right-hand side is the total energy radiated from the body, assuming blackbody radiation ($\sigma = 5.67 \times 10^{-8}\ \mathrm{Wm^{-2}K^{-4}}$ is the Stefan–Boltzmann constant). The second term is energy absorbed by sublimation of ice, L being the latent heat of sublimation (2500 kJ $\mathrm{kg^{-1}}$ for water ice at 273 K) and Z the sublimation rate in kg $\mathrm{s^{-1}\ m^{-2}}$. η is a geometrical factor, equal to 1 for a surface constantly perpendicular to the Sun, and 4 for a rapidly rotating body.

When there is no ice, or in a regime with negligible sublimation, we obtain

$$T = 278\,(1-A)^{1/4} r_h^{-1/2} .$$

This is the temperature predicted for an asteroid, and also for a cometary nucleus a long way from the Sun, before sublimation can occur. For small bodies (asteroids or satellites), temperatures around 273 K are observed at the distance of the Earth, and 120 K at the distance of Jupiter, in rough agreement with this rule. For planets like Earth, which have an atmosphere, temperatures may be significantly higher, owing to the greenhouse effect, which reduces re-emitted energy.

When ice is present, there is a strong coupling between sublimation rate Z and ice temperature. This is due to dependence of saturated vapour pressure on temperature. Close to the Sun ($r_h < 3$ AU), almost all received energy goes into ice sublimation. A thermostatic effect occurs, maintaining the temperature at about 180 K, regardless of the Sun's distance. The sublimation rate, however, is proportional to r_h^{-2}. Its maximum value at 1 AU may reach 0.5 g $\mathrm{s^{-1}\ m^{-2}}$, representing a global production of $Q[H_2O] = 10^{29}$ molecules $\mathrm{s^{-1}}$ over an area of 6 $\mathrm{km^2}$ of exposed ice. This would be the typical production rate for a reasonably active comet.

The rate of sublimation of ice is therefore a function of distance from the Sun (see Fig. 7.1). Far from the Sun, sublimation has little effect on equilibrium temperature. Below 150 K, which corresponds to a heliocentric distance of about 4 AU, sublimation is practically non-existent. The simplistic model for sublimation of pure water ice described above cannot therefore explain activity in distant comets.

A better understanding is obtained by re-examining some of the approximations made. For example, we neglected heat exchange with the inner nucleus. This would lead to a certain thermal inertia. Furthermore, cometary nuclei are not homogeneous. Active ice regions only occupy a small part of their surface. Variations will result from rotation of the comet (day–night effect), and they will also depend on orientation of the rotation axis relative to the Sun (seasonal effect). These considerations explain why the observed variation of $Q[H_2O]$ is only very roughly proportional to r_h^2.

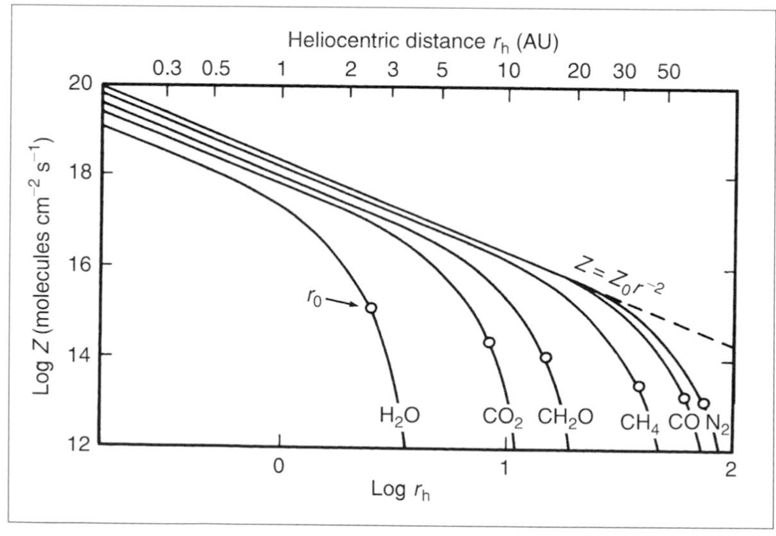

Figure 7.1.
Sublimation rates for ices of H_2O, CO_2, CO, and several other molecules as a function of heliocentric distance. Water effectively ceases to sublime beyond 3 AU, whereas CO and N_2 may be responsible for cometary activity at far greater heliocentric distances. Adapted from Delsemme, 1982 [41].

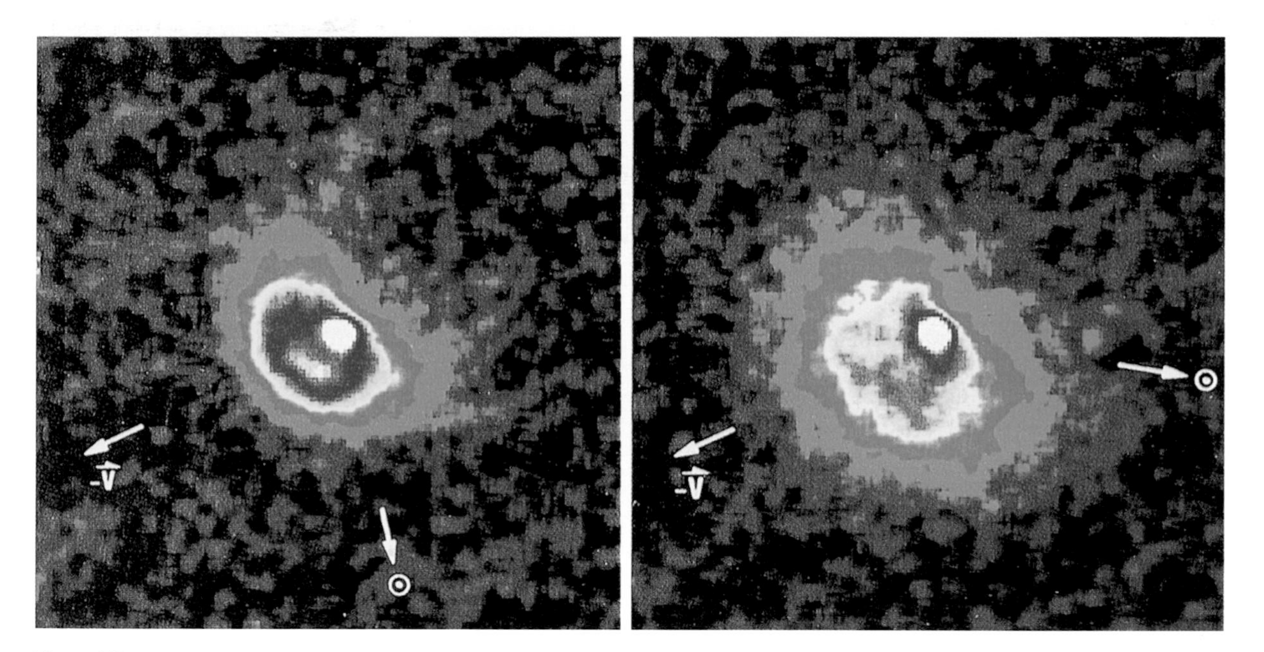

Figure 7.2.
Distant activity of comet Halley. Two stages in a burst of activity in 1991, on 12 February (*left*) and 12 March (*right*). The comet was at 14 AU from the Sun. Photo taken at ESO, La Silla, with the Danish 1.54 m telescope. Courtesy of ESO.

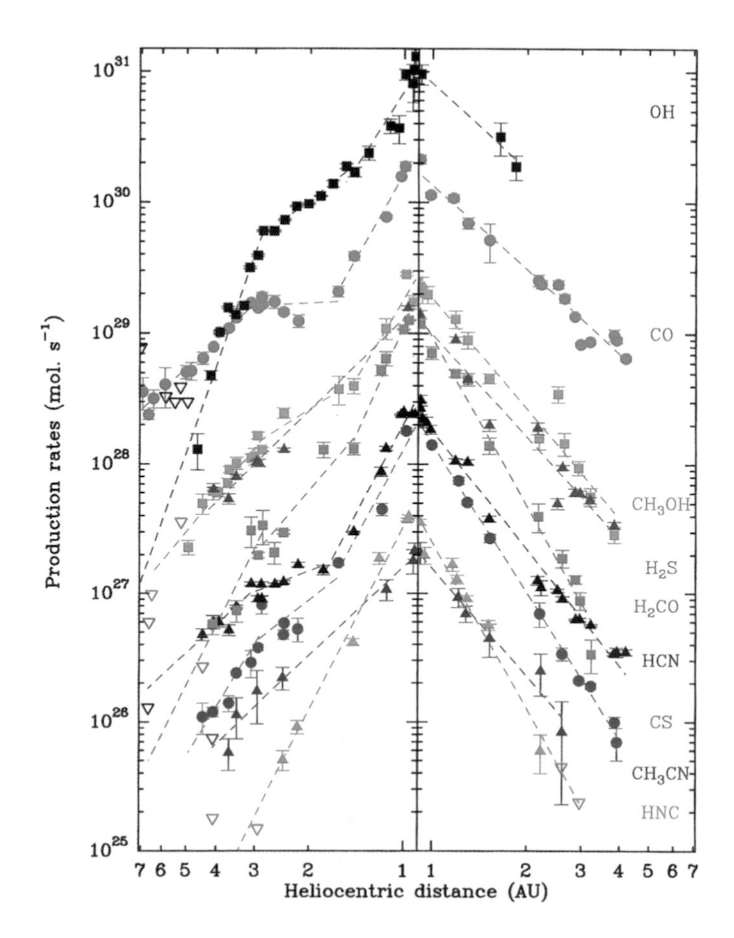

Figure 7.3.
Evolution of molecular production rates as a function of heliocentric distance, established from radio observations of comet Hale–Bopp spanning two and a half years. Adapted from Biver *et al.*, 1998 [28].

7.2 Coma formation and jet hydrodynamics

An important part of cometary physics is concerned with understanding the flow of gas and dust released by the nucleus. Close to the nucleus, gas molecules are subject to frequent collisions. The gas is then governed by the laws of classical hydrodynamics. Assuming spherical symmetry, flow is radial with speed v_{exp}. In Chapter 4, we saw that gas density n at distance r is given by

$$n(r) = \frac{Q}{4\pi r^2 v_{\text{exp}}} ,$$

where Q is the molecular production rate, i.e. the number of molecules leaving the nucleus per unit time.

The density thus decreases rapidly with expansion of the gas. Hence, the molecular mean free path (i.e. the mean distance a molecule travels between two consecutive collisions) grows longer and longer. Far from the nucleus, collisions become negligible. This is the free molecular flow regime, and it is no longer possible to apply formulas from classical hydrodynamics. Calculations can nevertheless be made by simulating paths of individual molecules. Their change of trajectory at each collision is simulated randomly, and the mean taken over a large number of particles. This is called the Monte Carlo method. It requires a great deal of computer time to perform such computations, but results are good.

This flow occurs in the form of supersonic jets, similar to those used for aeronautic propulsion and in certain laboratory experiments. The result is an adiabatic expansion, in which no energy is exchanged with the environment. The pressure, governed by sublimation conditions close to the icy surface, can thereby continuously transform to the almost zero value which exists in interplanetary space. The expansion is accompanied by a sudden and significant temperature drop, together with an acceleration (see Fig. 7.4).

In reality, expansion is not strictly adiabatic and cometary gases do undergo some energy exchange with the environment. Firstly, they radiate, emitting rotational and vibrational lines, which tends to cool them. Secondly, during photolytic processes (see Section 4.4) which destroy molecules, fragments are ejected at high speed. This tends to heat the gas by collisions. In fact, these processes have only limited efficiency. Heat losses by radiation are compensated

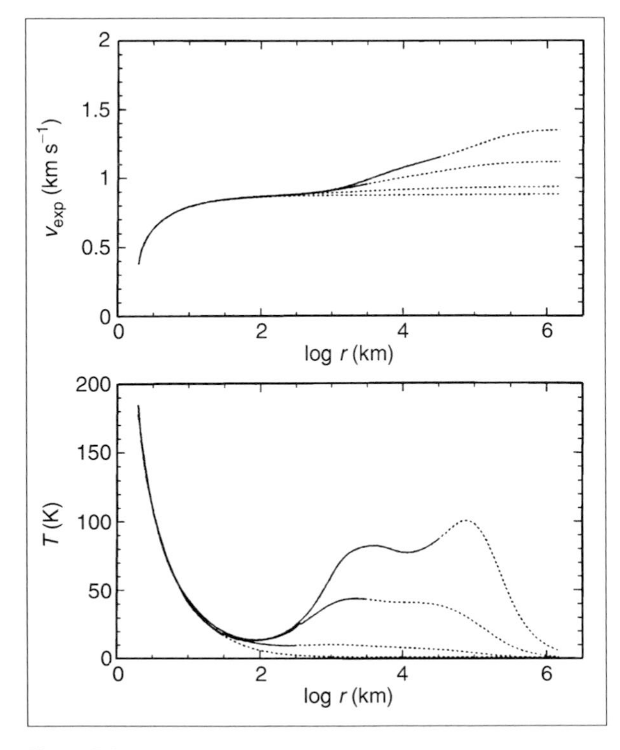

Figure 7.4.
Variation of expansion speed (*upper*) and temperature (*lower*) in a cometary atmosphere as a function of distance from the nucleus. The various curves, calculated from hydrodynamic models, correspond to production rates of 10^{27}, 10^{28}, 10^{29} and 10^{30} water molecules per second (speed and temperature values increasing with production rate). *Dotted curves* denote regions at the limit of validity for classical hydrodynamics, because collisions are not frequent enough.

in denser regions of the atmosphere by reabsorption of this same radiation. Photolytic heating is only efficient if fast hydrogen atoms produced after the destruction of H_2O can be slowed down by a series of collisions. This is not the case in the outer coma where densities are too low.

The overall result remains uncertain. We would nevertheless predict that, in large and prolific gas-producing comets with an extensive collisional region, photolytic heating will dominate over cooling due to expansion. Temperatures and speeds will therefore increase, as shown in Fig. 7.4, and the effect will be all the more evident as the comet approaches the Sun, because photolysis rates are proportional to r_h^{-2}.

What has actually been observed? Expansion speeds of cometary atmospheres can be estimated from molecular radio line profiles (see Chapter 6). Speeds of the order of 0.8 km/s have been deduced for small comets at 1 AU from the Sun. These speeds increase as predicted by theory when the comet approaches the Sun, and when its molecular production rate is higher. For comet Hale–Bopp, the expansion speed was observed to increase from 0.5 to 1.2 km/s as the comet approached from 6 to 0.9 AU. The Giotto probe has also measured the increase in expansion speed with distance from the nucleus, by directly measuring molecular energies when it passed through comet Halley's atmosphere.

Temperature measurements are more delicate. However, an assessment can be based on relative intensities of rotational or rotational-vibrational molecular lines (see Fig. 6.13 in Chapter 6). Temperatures deduced in such a manner do not really correspond to the gas temperature unless there are enough collisions to ensure thermal equilibrium. In practice, this is rarely the case. Temperatures based on rotational lines of methanol, in the radio region, and of water, in the infrared, typically lie in the range 30–60 K for small comets at 1 AU from the Sun. These are higher than simple models would predict. It is likely that other effects come into play, such as heating by dust, or recondensation of water into mini-droplets, which might oppose excessive cooling. For comet Hale–Bopp, the temperature was observed to increase from about 10 K at 6 AU to more than 110 K at perihelion, 0.9 AU from the Sun.

Interactions between gas and dust should also be taken into account. In the collision region, gases exert pressure on dust grains which gradually accelerates them. Once outside this zone, the grains decouple from the gas, conserving the terminal velocity acquired in the previous stage. This velocity depends on grain size. Thus small grains can attain the gas expansion speed, but larger ones may not even reach escape velocity, remaining captive in the cometary gravitational field.

7.3 Break-up of cometary nuclei

There are many different ways in which a comet may evolve. Some of these could be considered as 'natural'. For example, there may be a gradual change in the level of activity with varying distance from the Sun; there may be seasonal effects due to tilting of the rotation axis, and the occasional burst of activity as dust shells evolve. On the other hand, certain 'catastrophic' events are known to happen, involving break-up of the nucleus or collisions with other objects in the Solar System.

7.3.1 Break-up under effects of cometary activity: comets P/Biela and West

A historic example is provided by comet 3D/Biela. This was the third comet, after 1P/Halley and 2P/Encke, to have its periodicity proved and determined (6.9 years), by Jean

Gambart at the Paris Observatory. It was discovered in 1826 by Austrian astronomer Wilheilm von Biela (1782–1856), and associated with previous cometary apparitions in 1772 and 1805. When it reappeared in 1832, there were fears of a collision with Earth. In its 1846 passage, two pieces were observed, each with a tail, and they were reobserved at the following passage in 1852. Since then, this comet has never been seen again. Among recent examples of cometary break-ups, we may cite the gorgeous comet West (1976 VI), whose nucleus split into four fragments soon after its passage at perihelion, and comet 73P/Schwassmann–Wachmann 3

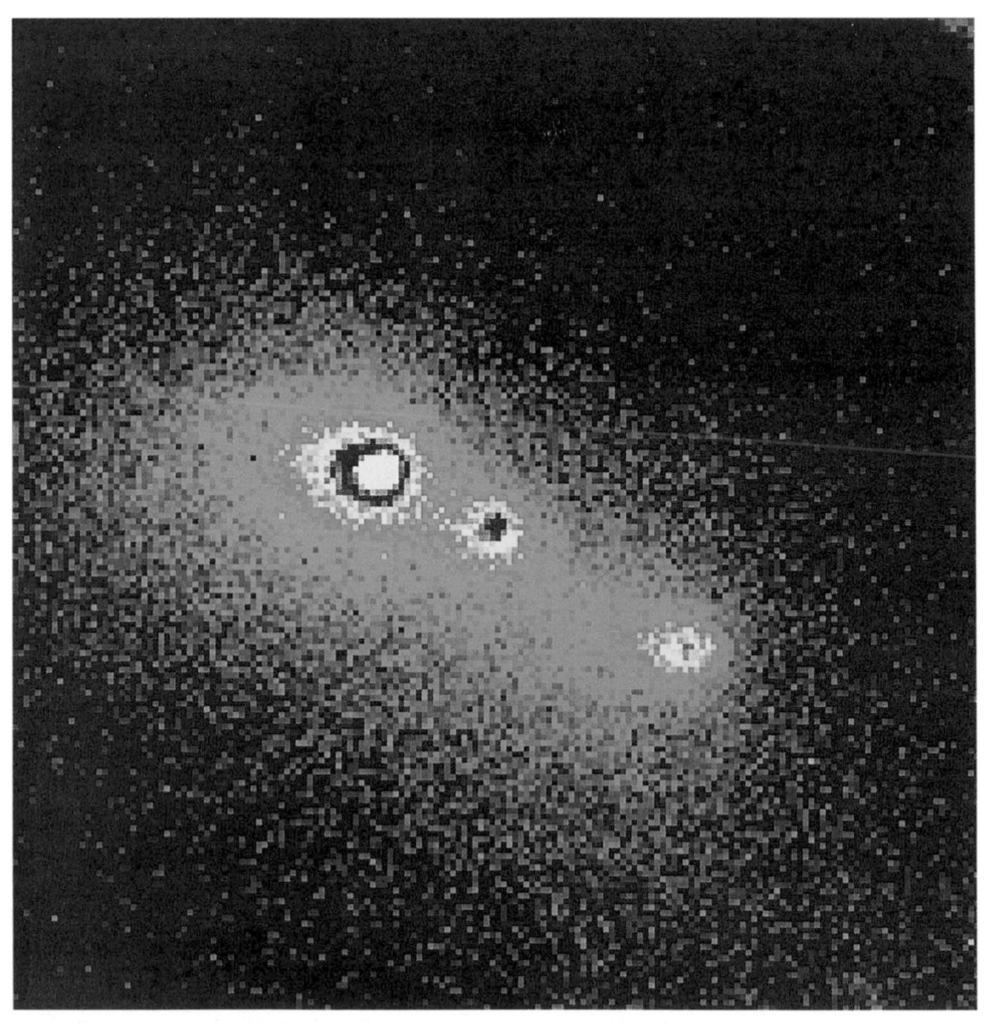

Figure 7.5.
In this image taken with the 3.5 m New Technology Telescope at ESO on 31 January 1996, comet 73P/Schwassmann–Wachmann 3 appears as three condensations, each with its own tail. This Jupiter-family comet, discovered in 1930, underwent a major outburst, first observed as an increase in OH production at the Nançay radio telescope in September 1995. It then brightened and was observed by many amateurs. As shown here, the outburst was accompanied by a splitting of the nucleus into at least three parts. Outer fragments are separated by 17 arcsec, which corresponds to 31 000 km at the comet, and are expanding apart with a velocity of about 6 m/s. ESO document.

(Fig. 7.5). This type of event is not uncommon. More than twenty cases have been recorded, corresponding to about 2% of perihelion passages. Nuclear break-up must therefore be considered as a possible outcome in any comet's evolution.

In some cases, such as comet Hyakutake in March 1996, we have witnessed the separation of one or more small fragments of nucleus. These fragments have then become inactive and invisible to observation. In other cases, a nucleus has burst into several fragments of roughly equal size. Such fragments separate slowly (at speeds of a few metres per second). Fragmentation is accompanied by an overall increase in brightness, for two reasons. Firstly, a greater area is exposed to the Sun, and secondly, fresh ice is revealed within the nucleus. The increased brightness is only temporary, particularly in the case of small fragments, which seem to be used up rather quickly.

Various phenomena may trigger nuclear break-up, but the main cause is the extremely low cohesion in these objects. This fragility could result from the way cometary nuclei are formed (see Fig. 7.6). They seem to be built up from mini-nuclei, or planetesimals, coming together in low speed collisions; at the slightest hint of internal stresses, they tend to come apart once again.

7.3.2 Break-up under tidal forces: the Roche limit

If one celestial body passes close enough to another, it may break up under the effect of

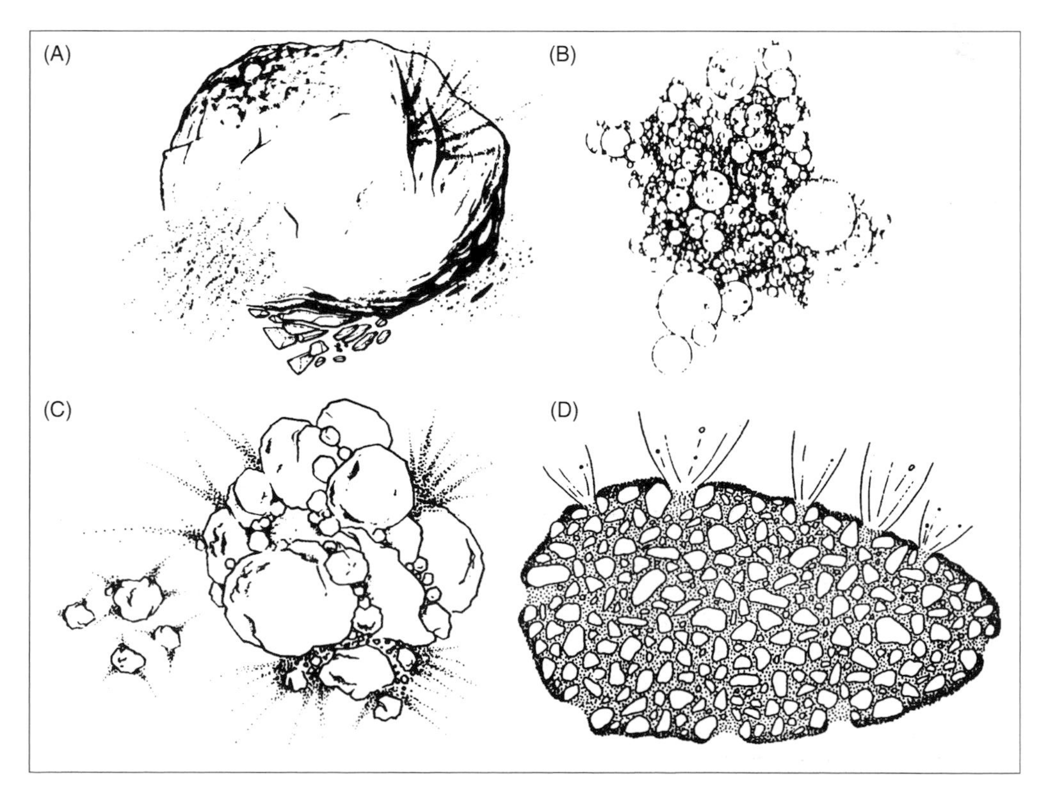

Figure 7.6.
Different models for the structure of cometary nuclei. (A) Whipple's icy conglomerate. (B) An aggregate of flakes, with fractal structure. (C) Accumulation of primitive debris. (D) Composite model in which refractory blocks of different sizes are bound together by ice. From Donn, 1991 [44].

tidal forces. Such forces are due to a differential effect of gravity. Parts of the satellite close to the source of gravity are attracted more strongly than distant parts, resulting in a stretching effect. Tidal forces are opposed by the satellite's own gravity, and by its cohesion if it is a solid. In the last century, Edouard Roche (1820–1883) showed that, for a liquid satellite in orbit about a star, break-up occurs when its orbital radius goes below a certain critical value r_R. This is a function of the star's radius and density, r_0 and ρ_0, respectively, and the satellite density ρ_s:

$$r_R = 2.44 r_0 \left(\frac{\rho_0}{\rho_s}\right)^{1/3} .$$

The *Roche limit* r_R is difficult to estimate for comets. For one thing, we do not know their density, although it is probably close to 0.5, and certainly less than the density of Jupiter (1.31) or the Sun (1.41). For another, the cometary nucleus is not liquid, having a certain degree of cohesion. This cohesion is very weak, but it opposes break-up and will tend to decrease r_R.

Jupiter plays an all important role with regard to the orbits of short period comets (see Chapter 2). Several comets are known to have become temporary satellites of Jupiter. Recent examples are comets P/Gehrels 3 and P/Helin–Roman–Crockett. Such orbits are subject to influences from the Earth and the Sun, and are not stable. After a few revolutions around the planet (between one and four, rarely more), the comet escapes from Jupiter. Along this part of its trajectory, the comet may pass close enough to Jupiter for its nucleus to break, or it may even crash into the planet. This happened to comet Shoemaker–Levy 9 in July 1994 (see Chapter 8).

Comet P/Brooks 2 appeared in 1889 with five nuclei. It was established that it had passed three years previously at only two Jovian radii from Jupiter's centre, and that break-up had occurred then.

7.3.3 Sungrazing comets

Several comets have passed very close to the Sun, at less than two solar radii. Some of these events were truly remarkable, notably those of 1680, 1843 I, 1882 II, 1887 I, and more recently, Ikeya–Seki 1965 VIII (see Appendix, Table A.1). Others such as 1945 VII and 1963 V were less noticeable. Still others, much smaller, were difficult to see at all, and specially adapted instrumentation was required to observe them. Fifteen comets were discovered in this way by SOLWIND between 1979 and 1984, and by SMM (Solar Maximum Mission) between 1984 and 1989. The LASCO instrument on SOHO discovered no fewer than sixty comets between 1996 and 1998. SOLWIND, SMM and SOHO were satellites devoted to solar observation and were equipped with coronagraphs (telescopes designed to create eclipse conditions artificially by masking the bright solar disk). It was thus possible to explore the immediate vicinity of the Sun without being dazzled. (The SOLWIND, SMM and SOHO comets are all very small objects. They were only detectable because of their close approach to the Sun. Although approximate orbits could be determined, these comets were not seen again after the satellite observations.)

The nuclei of some of these comets were observed to split following perihelion passage. Their activity generally diminished rapidly after perihelion. Their tails remained, however, even after the coma had disappeared. Other comets do not survive perihelion at all. In these cases, at least for the smaller comets, total vaporisation causes the comet to disappear altogether. It is then more difficult to observe any splitting of the nucleus than it was for P/Shoemaker–

Figure 7.7.
On 1 June 1998, the LASCO coronagraph aboard the SOHO satellite observed two comets plunging towards the Sun. They can be seen with their tails on the bottom right part of the image. The white circle indicates the solar disk. The larger, dark disk is the region occulted by the coronagraph. The radial structures are jets of matter in the solar corona. Courtesy ESA/NASA.

Levy 9. According to a hypothesis of Heinrich Karl Kreutz (1854–1907), formulated over a century ago and refined by B. Marsden on the basis of more recent observations, all of these comets may originate from the same body. This object is assumed to have a period of several centuries, and to have gradually fragmented during successive passages close to the Sun. The sungrazers are thus also known as the *Kreutz family*. Under the influence of planetary perturbations, in particular from Jupiter, orbits and orbital periods of the various fragments would be modified in different ways. They should return to almost the same perihelion, just as we observe, but at different times. We are therefore witnessing the same type of interaction as that between comet Shoemaker–Levy 9 and Jupiter, but on a much larger scale. In that case, the nucleus broke up

Figure 7.8.
An experiment carried out by J.M. Greenberg. A mixture of water, methane and ammonia ices was deposited in a cryogenic container, then subjected to UV radiation. The yellow substance formed was a complex mixture of organic molecules, considerably enriched in carbon. Photos show the sample before irradiation (*left*) and after (*right*). From Greenberg, 1984 [47].

Artificial comets and laboratory simulations

Attempts have been made to produce artificial comets by releasing barium gas in the upper atmosphere. Ionisation, fluorescence and diffusion were then observed. An unintentional experiment was observed from the European Southern Observatory in Chile, in January 1992. A UFO was photographed and later identified as a block of ice from the space shuttle air conditioning system, which had been dumped in space. However, these phenomena occurred in the upper atmosphere, and not in interplanetary space. They are thus more relevant to terrestrial aeronomy than to cometary physics.

Many laboratory experiments have reproduced cometary processes. Certain parameters important in understanding comets can be observed and measured at first hand. A laboratory technique which has now become commonplace is the simulation of supersonic molecular jets. This reproduces some of the physical conditions in cometary atmospheres, e.g. adiabatic expansion in a vacuum, where temperatures can fall to a few kelvins. It also allows observation of molecular spectra in conditions very similar to those prevailing in comets.

Systematic studies have been made in several laboratories of very low temperature ice mixtures, to simulate cometary conditions, and also conditions on the outer planets and their satellites, and interstellar grains: the experiment by Mayo Greenberg and coworkers at Leyden (Netherlands) is shown in Fig. 7.8, and others were carried out at the glaciology laboratory, Université de Grenoble (France) and the NASA Ames Center, California (USA). In this way, standard reference spectra are obtained for ice mixtures; retention capacities of ices for volatile gases are measured; fractional sublimation of ices is studied to determine its temperature dependence; and finally, chemical reactions, in particular the synthesis of complex molecules, can be observed when ice mixtures are irradiated with UV radiation or ions.

More sophisticated simulation experiments have been carried out for the KOSI project (Kometen Simulation) in Cologne (Germany), and likewise in the former USSR. A large sample of a mixture of ices and mineral dusts was placed in a container of diameter 130 cm. This was then introduced into a space simulator, where temperature and vacuum are controlled, and subjected to a simulation of the Sun's radiation. Probes, cameras and spectrometers measured temperature variations in the sample, photographed and filmed matter ejection, and analysed the composition of sublimed material. The experiment continued over several days. At the end, remains of the sample were chemically and physically analysed.

These experiments provide some of the parameters needed to model cometary phenomena, to interpret observations and to carry out calibrations. They also prepare the ground for *in situ* analysis of cometary material in future space missions, not to mention analysis of cometary samples we hope, in the longer term, to bring back to Earth.

in July 1992. The fragments came back and struck the planet in July 1994, at intervals of about one week.

All these observations demonstrate that cometary nuclei are very fragile bodies. But how fragile are they? This is measured by the so-called *tensile strength* parameter, which has dimensions of pressure. Evaluations of this strength lead to a wide range of values, from 10^2 to 10^4 Pa (1 Pa (pascal) is close to 10^{-5} kg/cm^2). These are very low values, testifying to the porous nature of cometary material. For comparison, the tensile strength of compact material is 10^6 Pa for pure water ice, one or two orders of magnitude greater for rock and stone, and 4×10^8 Pa for iron. The truth about tensile strengths of cometary nuclei will ultimately be revealed by space probes, when they attempt somewhat risky landings on these bodies.

8

The collision of comet Shoemaker–Levy 9 and Jupiter

In July 1994, a comet crashed into Jupiter. It had already broken into more than twenty fragments before the collision, and a series of impacts ensued between 16 and 22 July. This event had been predicted more than a year before. It was carefully followed by a vast campaign which mobilised practically every available means of astronomical observation.

We were thus able to observe explosions of an energy never before witnessed in the Solar System, and also to gather exceptional data concerning the atmosphere of Jupiter and its reaction to this significant perturbation. In the present chapter, however, we shall be interested in the nature and composition of the body which struck Jupiter. A great deal can be deduced from the information obtained.

8.1 Background to the discovery

Carolyn and Eugene Shoemaker, with the help of David Levy, were involved in an asteroid research programme over a number of years. They used the 40 cm Schmidt telescope at the Mount Palomar Observatory to expose photographic plates, observing the antisolar region around the time of the new Moon. Each year, this programme discovered several dozen new asteroids, as well as a few comets. In all, thirteen comets, of which nine are short period, carry the name Shoemaker–Levy.

Examining a plate exposed on 24 March 1993, they discovered a comet which was at once recognised to be unusual. It appeared to be rather bar-shaped or, in C. Shoemaker's own terms, like a 'squashed comet'.

More accurately produced images later revealed that it was made up of more than twenty almost perfectly aligned condensations, 'like pearls on a necklace' in the words of David Jewitt and Jane Luu, each one with its own tail (see Fig. 8.1). The comet was first called Shoemaker–Levy 1993e, then renamed P/Shoemaker–Levy 9 when it had been established that it was a short period object. Its official designation in the new nomenclature is D/1993 F2, where D indicates that it has disappeared. In the following, we shall use the abbreviation SL9.

Orbital calculations showed that SL9 was in orbit around Jupiter with a period of about two years and that it had passed at a distance of only 1.6 Jovian radii from the planet on 8 July 1992. It was deduced that the comet must have broken up during this close passage, under the influence of tidal forces, just as comet P/Brooks 2 had fragmented in 1886 when it likewise passed very near to Jupiter.

Extrapolating calculations further into the past, it could be deduced that SL9 was

Figure 8.1.
Nuclei of comet Shoemaker–Levy 9 observed by the Hubble Space Telescope on 17 May 1994. The first and last fragments, labelled A and W, were then separated by 360 arcsec, implying a distance of 1.15×10^6 km in real terms. From Weaver *et al.*, 1995 [80].

captured by the planet at some time between 1920 and 1938, and that prior to this capture, it was probably a short period comet in the Jupiter family, rather than a comet coming directly from the Oort cloud or the Kuiper belt. Assuming this to be the case, SL9 must have been orbiting Jupiter for several decades before it was noticed.

The orbital calculations also showed that SL9 would reach its next 'peri-Jove' in July of 1994, and that it would pass at only 0.6 Jovian radii from the centre of the planet. In other words, a collision was foreseen.

8.2 The comet before impact

Fragments were labelled A, B, up to W, in the order they were to collide with Jupiter. They were all active, with the exception of I and O. Over the fifteen month period in which they were observed, certain disappeared (J and M), and others split (P into P_1 and P_2, Q into Q_1 and Q_2).

No gases were revealed by spectra of the various pieces; neither OH nor CN in the visible, nor CO in the radio region. Indeed, we have already seen that, at this distance from the Sun, sublimation of water ice is negligible. It may be, however, that some production of molecules such as CO or N_2 would have gone unnoticed. Even a very small production of gas (a few kg/s, which would be undetectable) would be enough to maintain dust production and hence generate observed levels of activity.

The nuclei spread out relative to each other, whilst remaining in line right through the final orbit (see Fig. 8.2). At the time of their discovery, they occupied a length of 150 000 km, compared with five million kilometres just before the first impacts. As numerical simulations have shown, this gradual separation happens naturally when fragments separate initially at the same instant with some small relative velocity (see Fig. 8.3).

Concerning the size of fragments, assessment is made difficult by the presence of dust, as in the case of any active comet. From image analysis, the larger fragments were estimated to be about one kilometre across, implying a diameter of two or three kilometres for the original comet. This agreed with models of break-up, made in order to understand how the train of fragments had evolved.

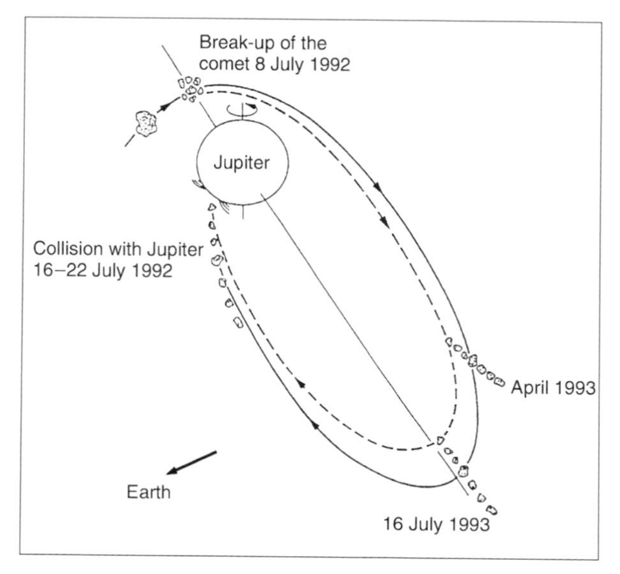

Figure 8.2.
The changing appearance of comet Shoemaker–Levy 9 (1993e) between 1992 and 1994, during its last orbit around Jupiter. Courtesy of the Société astronomique de France.

Taking into account uncertainties relating to their sizes (0.5 to 1 km) and their densities (0.2 to 2 g/cm^3), the fragment masses lay somewhere between 5×10^{10} kg and 4×10^{12} kg. As they were to strike Jupiter with a speed of 60 km/s, this meant collision energies would be within the range 4×10^{20} joules and 3×10^{22} joules, equivalent to somewhere between 8×10^4 and 6×10^6 megatonnes of TNT. For comparison, the atom bomb which totally destroyed Hiroshima was equivalent to a mere 20 kilotonnes.

8.3 Observation of impacts

It seemed that the impacts themselves would be difficult to observe, occurring as they would on the hemisphere of the planet which was hidden from terrestrial view (see Fig. 8.4). However, thanks to the planet's rotation about its axis, the impact zones came into sight about fifteen minutes after the event.

One observatory was nevertheless able to observe the impacts directly: the space probe Galileo, which was on its way to Jupiter at this time. Unfortunately the probe had suffered a communications break-down, as its large antenna had not functioned correctly. A transmission rate of only ten bits per second could be established, so observation was feasible, but data was only partially transmitted. Even then, it had to be spread out over a period of several weeks!

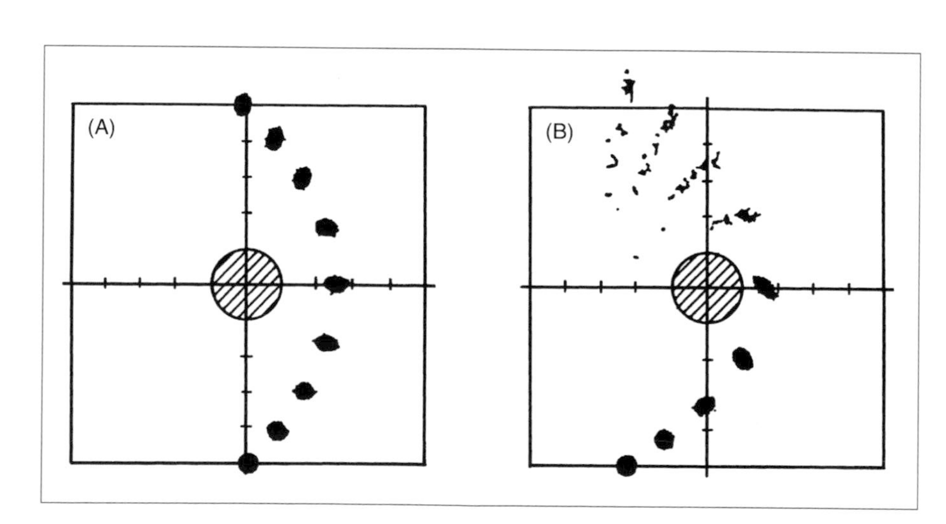

Figure 8.3.
Simulation of break-up for a small body passing close to a planet. On the left, the body does not pass inside the Roche limit and is merely deformed. On the right, it passes within the Roche limit and breaks up; fragments are lined up in a bar shape. This simulation was carried out by a Japanese team ten years before discovery of SL9. From Hayashi *et al.*, 1985 [51].

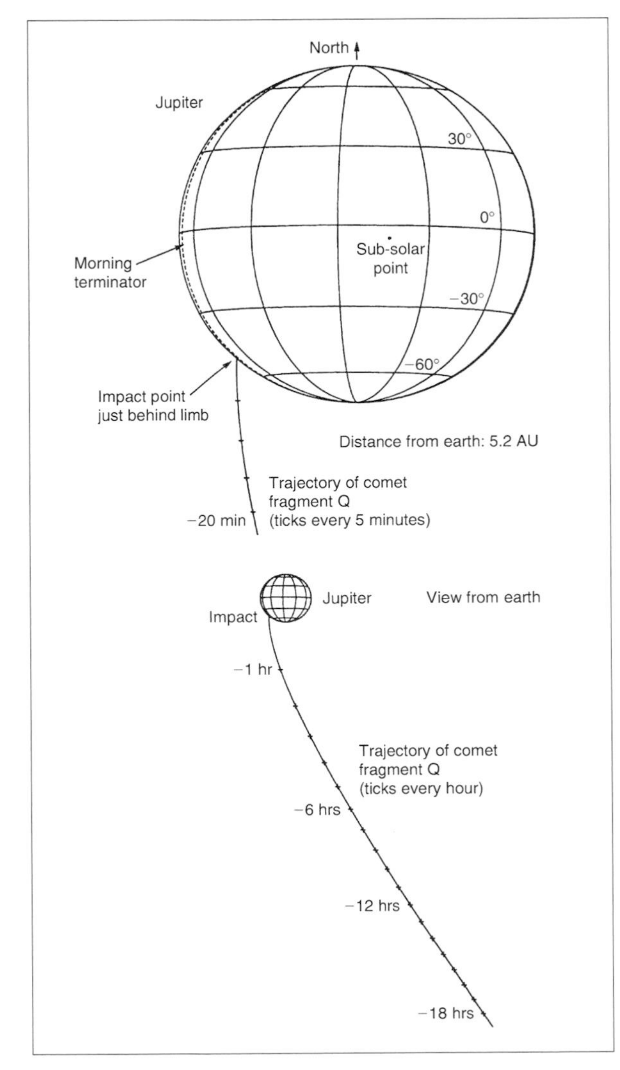

Figure 8.4.
Geometric configuration of the impacts of SL9 on Jupiter, as seen from Earth. All impacts occurred on the hidden face of the planet, but just behind the limb and close to the dawn terminator. Courtesy of the Jet Propulsion Laboratory, Caltech.

The phenomenon turned out to be truly spectacular, sometimes exceeding the most optimistic predictions. The scars left on Jupiter's surface were visible for a considerable time afterwards, even to modest telescopes of amateur astronomers. Almost all observation programmes led to good results.

The Galileo probe was in a position to detect the first moments of some impacts, although this was not known until later. In 'real time', infrared imaging techniques provided the best view of how impacts were evolving. Indeed, events hidden on the other side of the planet were clearly detectable through the immense release of heat, together with a huge plume which reached up to high altitudes. 'Craters' soon appeared (see Fig. 8.5), measuring about 10 000 km in diameter. These were formed by the fallout of matter, dust and aerosols, raised up during impacts.

The most violent collisions, which left the most noticeable marks, were not caused by the brightest pre-impact fragments. This suggests that fragments may have had differing densities and compositions, and that the cometary nucleus prior to fragmentation was heterogeneous.

Spectroscopic observations of explosions and impact sites revealed signals from several new or modified atomic and molecular species. Some of these effects only lasted for a few minutes after impact, whilst others lasted for days, months, or even years.

* In the UV region, the Hubble Space Telescope detected absorption bands of NH_3, CS_2, CS, S_2 and possibly H_2S, as well as emission lines of neutral and ionised atoms (hydrogen, magnesium, silicon and iron).
* Visible spectra observed from the ground (in particular at the Pic-du-Midi site in France) revealed an intense but transient emission, lasting for only a few minutes after impact, of metallic lines Li, Na, K, Ca, Mg and Fe. It is worth remembering that corresponding lines had been observed when comet Ikeya–Seki passed close to the Sun in 1965, although lithium had not been detected on that occasion. These lines are related to metallic lines

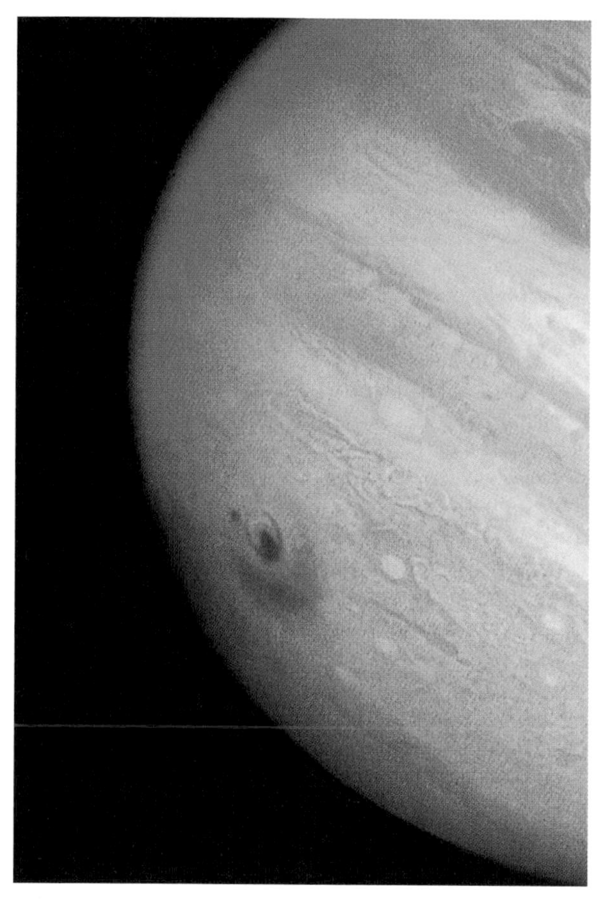

Figure 8.5.
The impact site of fragment G on Jupiter, 1 h 45 min after the collision. The dark central region, corresponding to the impact itself, measures about 2500 km in diameter. It is surrounded by a thin, dark ring, probably due to a wave generated by impact. Further out is a crescent shape created by the falling back of matter ejected during impact. Courtesy of the Hubble Space Telescope.

observed in the UV by the HST. It is not clear whether the emission mechanism might be fluorescence, an emission cascade following recombination of atomic ions created in the explosion, or possibly collisional excitation as disturbed matter fell back into Jupiter's atmosphere.

* Normally, Jupiter's infrared spectrum shows some absorption bands of methane, carbon monoxide and ammonia. After the impacts, these same bands appeared in emission, together with those of certain hydrocarbons, such as acetylene (C_2H_2) and ethane (C_2H_6). The shape of the bands corresponded to a temperature in the range 500 to 600 K, which decreased with a characteristic time of a few hours as the atmosphere cooled off. The stratospheric plane KAO observed water lines at 7.7 μm and 22 μm, and water bands were also observed in the infrared at around 2 μm by the Anglo-Australian Telescope. These were transient emissions, lasting only for a few minutes after the impacts, possibly because gases had by then cooled, rather than because the water had disappeared. Water was not otherwise observed during these events (recall that water is very difficult to observe from Earth).

* No radio line had ever been detected in Jupiter's atmosphere before this time. Observations were made using the IRAM 30 m radiotelescope, and the 15 m radiotelescopes of the JCMT and the SEST (Swedish-ESO Submillimetre Telescope in Chile), as well as the IRAM millimetre radio-interferometer on the Plateau de Bure in France. Lines due to carbon monosulphide (CS), carbon oxysulphide (OCS) and hydrogen cyanide (HCN) were all detected. These lines were observed in emission for several days after the impacts, with surprising intensity. They then switched over to absorption, remaining detectable for several months, or even years, in the case of HCN. This development indicates the major thermal changes which must have taken place at impact sites in the Jovian atmosphere.

8.4 Constitution of the object

What can be said about observations of these chemical species, and what can be deduced about the constitution of the body which struck Jupiter? During collisions, the temperature rose to over 10 000 K, sufficient not only to vaporise any matter present, but also to dissociate molecules and ionise atoms. It follows that any information concerning the object's chemical constitution would have been lost; only its elementary (atomic) composition could be retrieved. During rapid cooling of the 'fireball', atoms in the impactor would have recombined with atoms in the Jovian atmosphere, in a kind of 'shock chemistry', creating molecules which might be significantly different from any of the initial compounds. It may also be that matter was able to rise up from lower levels in the Jovian atmosphere, supposing that some fragments succeeded in penetrating that far. These layers are usually quite inaccessible to any means of observation, but their composition can be assumed very different from that of the upper layers. They are likely to be made up of clouds rich in water, ammonia, and sulphur compounds such as ammonium hydrogen sulphide (NH_4SH). In order to interpret observations, a good understanding of all these physical and chemical processes is essential.

However, part of the matter making up fragments may have been pulverised in successive steps, thus forming micro-fragments which would be slowed down without being completely vaporised; and constituent molecules of such micro-fragments might escape dissociation. Dark clouds of dust were indeed observed at impact sites, so that a small proportion of preserved molecules from fragments could have survived collision.

It is reasonable to assume that metals whose atomic lines were detected originated in fragments, since upper levels of the Jovian atmosphere contain no metallic compounds (these would sediment into lower layers). It may also be assumed that any oxygen, observed in compounds such as CO and H_2O, or sulphur, observed in compounds such as CS_2, CS, H_2S, OCS and S_2, would also have come from fragments. On the other hand, carbon and hydrogen in these molecules would very likely have been borrowed from H_2 and CH_4 molecules in the Jovian atmosphere. (The atmosphere of Jupiter is essentially made up of H_2 (89%), He (11%), CH_4 (0.17%) and NH_3 (0.02%).)

Our original question remains unanswered: what was SL9? Was it a comet or an asteroid? As we shall see in Chapter 10, this question may not be well posed, for the distinction between the two classes of object may not be clearly defined. In any case, the main information gleaned can be summarised as follows.

* SL9 was a small body which was captured by Jupiter from the inner Solar System. It was either a short period comet, an asteroid from the main asteroid belt, or a Trojan asteroid (see Chapter 10). This is not therefore a discriminating factor.

* SL9 was an extremely fragile body which broke up under Jupiter's tidal forces when passing at 1.6 Jovian radii. This would seem to confirm the low level of cohesion usually imputed to cometary nuclei.

* After break-up, fragments became a source of dust. No gases were detected, but this may just have been due to weakness of the fluorescence mechanism at such a great distance from the Sun. This kind of activity implies that the object was a comet, in the usual sense of the term.

* The mass, the size and hence the density of fragments remain uncertain. It is not yet possible to choose between the low density (0.1

to 1 g/cm^3) expected of a cometary nucleus and the somewhat higher density (over 2 g/cm^3) which would distinguish an asteroid.

* Water was identified by its infrared lines. Carbon monoxide and molecules containing sulphur were observed in significant quantities, quite compatible with those which would be contained in a large cometary nucleus composed mainly of ice. However, we have seen that observations of chemical species present after impact could not be used to draw firm conclusions. Any trace of molecules in the original impactor would have been lost in chemical reactions caused by shock of impact. Nevertheless, the large quantity of oxygen observed would seem to have originated in the projectile, since this element is lacking in upper layers of the Jovian atmosphere. However, it remains difficult to say whether it may have come from cometary ices (H_2O, CO, CO_2), silicates (SiO_4) in cometary dust, or even from asteroidal rocks.

On the whole, evidence suggests an object having little internal cohesion, therefore a comet, but it is not possible to make a complete assessment of the composition of the nucleus, or even to estimate the mass ratio of refractory dust to volatile ices.

The break-up of the nucleus of SL9 and its collision with Jupiter represent a remarkable set of events in several respects. If the comet had not broken up in a first stage, it would have remained unknown until the collision itself. Although the results of impacts would not have gone unnoticed, the first few moments of the event would certainly have been missed, and an observational campaign could not have been properly prepared. Moreover, the same phenomenon could be repeatedly observed, as the various fragments came one by one into collision. In this way, the sometimes surprising results of early observations could subsequently be confirmed.

8.5 Impact phenomena in the Solar System

It is natural to wonder how often such events might happen. Instrumental astronomy goes back only a little more than three centuries. Several comets have been observed in orbit around Jupiter, but only recently, for such objects are rather faint. Two of these have broken up, namely P/Brooks 2 and SL9. The latter is the only confirmed case of a comet actually falling onto Jupiter. Any other collision with a comet of like size would have left an easily detectable trace, even if the comet itself had remained unnoticed before impact. In fact, new spots have occasionally appeared on Jupiter, and also on Saturn, although quite unlike the marks left by SL9 and hence were attributed rather to meteorological phenomena. Another impact of a small body on Jupiter may have been observed three centuries ago. On 5 December 1690, Jean Dominique Cassini (1625–1712) discovered a new spot on Jupiter. (J.D. Cassini was then heading the newly founded Paris Observatory; he was the author of many studies on Jupiter and Saturn, including the discovery of the Great Red Spot on Jupiter.) In the following days and weeks, the spot evolved, increasing in size and spreading out in longitude, in the same way as the Shoemaker–Levy 9 impact spots in 1995 (see Fig. 8.6).

A frequency of one SL9-type event every three centuries would seem quite plausible, but it should be said that much smaller objects may enter into collisions more often, and their consequences would not necessarily be noticed. Generally speaking, collisions in the Solar System are a very common phenomenon. This is

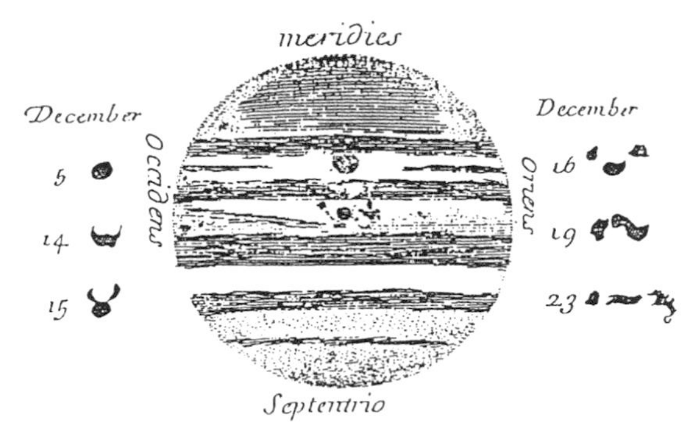

Figure 8.6.
Original drawing by J.D. Cassini, showing the apparition of a new spot on Jupiter in December 1690. Its evolution in the following weeks was strikingly similar to the spots created by the impacts of comet Shoemaker–Levy 9. From Cassini, *Nouvelles découvertes dans le globe de Jupiter*, 1692.

attested by the clear presence of many impact craters on the Moon, on other planetary satellites, and on the few asteroids that have been imaged by space probes.

8.6 Impact phenomena on the Earth

On a geological time scale, impacts of small bodies on Earth are not unusual, but their trace is often quickly removed by erosion. About two hundred cases have been recorded. Among these, the Meteor Crater in Arizona is undoubtedly the best known, with a diameter of 1.3 km and a depth of 200 m. It was created between 20 000 and 50 000 years ago by the impact of a small iron-rich asteroid, about 30 m in diameter, weighing around 60×10^6 kg, and travelling at 16 km/s. Its collision energy can be estimated at around 8×10^{15} J, or 2 megatonnes of TNT.

The so-called Tunguska event of 30 June 1908 left its mark over several dozen kilometres of the Siberian forest, although there was no crater. The object did not hit the ground, but exploded in the upper atmosphere, at an altitude of about 8 km, releasing an energy equivalent to about 10 megatonnes of TNT. The shock wave from this explosion was responsible for destroying the forests. It is supposed that the object had rather low density and was unable to survive penetration of Earth's atmosphere. A small stony asteroid, about a hundred metres across, or a comet would seem the most likely cause in this case.

Estimates of the frequency of meteorite impacts on Earth are shown in Fig. 8.7 as a function of their energy. Ten to twenty megatonne events, such as the Tunguska event, are not rare, occurring every few hundred years. Although their effects are catastrophic, they are nevertheless restricted to a limited area. By contrast, collisions with objects measuring more than one kilometre across could release energies exceeding a million megatonnes and are thus rather more worrying, for they could cause destruction on a global scale. Dust clouds rising from the impact and ash from forest fires would quickly reach the upper atmosphere and diffuse right round the Earth, thereby leading to long term climatic changes. Such an event would probably lead to mass extinctions of flora and fauna.

Disasters on this global scale might occur every few million years. Indeed, the larger of these events could explain the disappearance of some species, as observed by palaeontologists. It is thought that such an event may have happened about 65 million years ago, marking the transition between the Cretaceous and Tertiary periods. Palaeontologists have noticed that this date coincides with the disappearance of many species, including the dinosaurs. Geologists have discovered a sedimentary layer, dating from this same time, which is unusually rich in iridium, an element more commonly found on meteorites than on Earth.

The frequency of cometary and asteroidal collisions with Earth must have been much higher

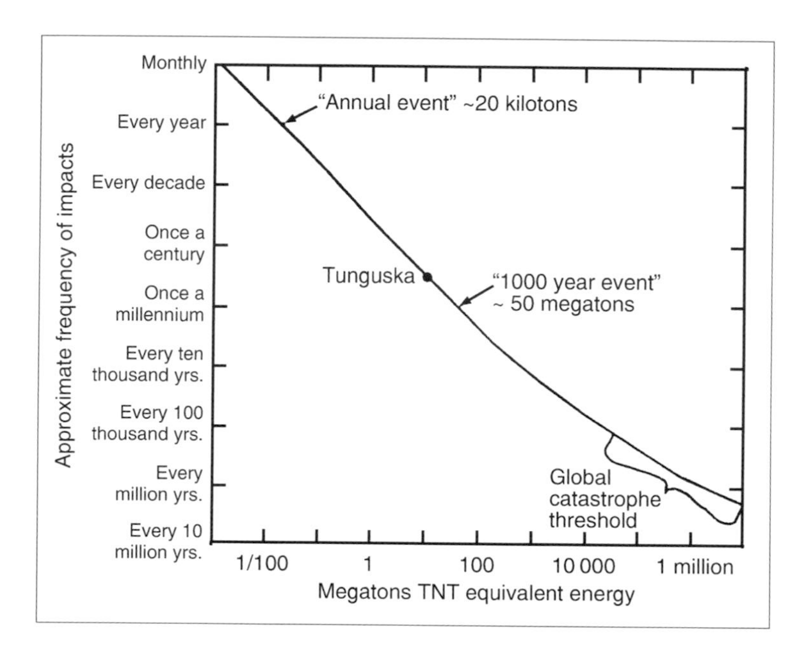

Figure 8.7.
Estimated frequencies of cometary and asteroidal collisions with Earth as a function of collision energy. From Morrison, 1992 [67].

during the few hundred million years following formation of our planet. Indeed, the latter would have been literally worked into shape by these collisions. According to some theories, the very oceans themselves might have been formed, at least partly, from cometary ices carried here by such collisions. According to other theories, prebiotic molecules carried by comets (we know of hydrogen cyanide, hydrogen sulphide, cyanoacetylene and formaldehyde, but others as yet undetected are likely to be present) may have played a major role in the appearance of life on Earth. If the latter scenario is accurate, a certain proportion of these molecules would have to be able to survive extreme conditions created by impact.

9
The nature of comets

9.1 Composition of cometary ices

We are now in a position to assess the composition of cometary ices (see Table 9.1).

* The main hypotheses of the Whipple model have been confirmed. Water predominates in cometary ices. This justifies the usual practice of giving molecular abundances with respect to water. However, we should make one reservation. Abundances given in Table 9.1 were measured in comets active at about 1 AU from the Sun, and we must take into account the phenomenon of fractional sublimation when deducing ice abundances in the nucleus from gas abundances in the coma. Models of the internal structure of comets would suggest that relative abundances in the coma may not be representative of their nuclear composition. These comets may well have lost a large part of their more volatile ices (e.g. CO, CO_2, N_2) before reaching the innermost regions of the Solar System. This idea is confirmed by observations of distant activity.

* Comets seem to be deficient in very volatile elements and molecules, like molecular hydrogen, noble gases and molecular nitrogen. Either these species were not able to condense during formation of cometary nuclei, or they were not retained in the nuclei because temperatures were not low enough.

* Ammonia and methane have long been suspected to be major cometary molecules (since the idea was put forward in the 1950s by Fred Whipple and Armand Delsemme). They were finally identified in comets Hyakutake and Hale–Bopp, but their abundances, of the order of 1%, are smaller than previously expected.

* In contrast, carbon monoxide, carbon dioxide and methanol are major constituents with abundances which may exceed 5%. Formaldehyde is also abundant, but its real origin – secondary source rather than nucleus – is still unclear.

* A wealth of other 'CHO' species have been identified: formic acid, acetaldehyde, methyl formate. Abundances are small, but they imply that many more species are probably there. This would suggest that the composition of comets is very complex. The presence in significant abundance of acetylene and ethane also suggests that a rich variety of hydrocarbons is present. All these species – those already identified and those still to be found – could well account for the formerly mysterious emission observed around 3.4 μm.

Table 9.1. Parent molecules identified in comets and their abundances.

Molecule		Abundance	Means of observation	Comments
H_2O	Water	= 100	IR	
CO	Carbon monoxide	2–20	UV, radio, IR	Extended source?
CO_2	Carbon dioxide	2–6	IR	
CH_4	Methane	0.6	IR	
C_2H_6	Ethane	0.3	IR	
C_2H_2	Acetylene	0.1	IR	
H_2CO	Formaldehyde	0.05–4	Radio	Extended source
CH_3OH	Methanol	1–7	Radio, IR	
$HCOOH$	Formic acid	0.1	Radio	
$HNCO$	Isocyanic acid	0.07	Radio	
NH_2CHO	Formamide	0.01	Radio	
CH_3CHO	Acetaldehyde		Radio	
$HCOOCH_3$	Methyl formate	0.1	Radio	
NH_3	Ammonia	0.5	Radio, IR	
HCN	Hydrogen cyanide	0.1–0.2	Radio, IR	
HNC	Hydrogen isocyanide	0.01	Radio	
CH_3CN	Methyl cyanide	0.02	Radio	
HC_3N	Cyanoacetylene	0.02	Radio	
N_2	Dinitrogen	0.02–0.2	Visible	Indirect, from N_2^+
H_2S	Hydrogen sulphide	0.3–1.5	Radio	
H_2CS	Thioformaldehyde	0.02	Radio	
CS_2	Carbon disulphide	0.1	UV, radio	Indirect, from CS
OCS	Carbonyl sulphide	0.4	Radio, IR	
SO_2	Sulphur dioxide	~ 0.2	Radio	
S_2	Disulphur	0.05	UV	

* The C_2 and C_3 radicals were identified very early on, in the visual spectra of comets. Their origin has long been a puzzle: they were *orphan molecules*. The recent discovery of acetylene (C_2H_2) and ethane (C_2H_6) can now fully account for the presence of C_2. There is also little doubt that the multi-step photodissociation of more complex hydrocarbons or carbon-chain molecules could explain how C_3 came to be there.

* There are many cyanides (molecules containing the CN group) which are parents for the CN radical. Among them, hydrogen cyanide (HCN) is the most abundant. It must be noted that the CN radical has been observed in some cases to have a partially jet-like distribution. Some people have attributed this to a direct production of CN from the organic mantle of cometary grains.

* Sulphurous molecules are also important. The most abundant is hydrogen sulphide (H_2S) which may exceed 1% in some comets, followed by CS_2 (traced by the CS radical) and SO_2.

It is true to say that we still have only a very rough overall understanding of cometary composition. It would seem to be very rich and complex.

For the moment only the simplest molecules have been identified. The presence of heavier and more complex molecules is confirmed by the infrared band at 3.28 μm, characteristic of aromatic molecules. It is also attested by significant signals corresponding to atomic masses above 50, which mass spectrometers recorded in P/Halley. It is quite probable that these signals come from a mixture of many molecules each occurring with low abundance. If so, they are likely to be difficult to identify. It is clear that we must await suitably equipped *in situ* analyses to resolve this problem (e.g. the Rosetta project, Chapter 11).

There is some debate as to the true origins of certain parent molecules observed. It has been established that molecules like CO and H_2CO do not all come from the nucleus. In fact they originate in an *extended source* in a circumnuclear region stretching out for several thousand, or even several tens of thousands of kilometres. Of course, CO and H_2CO may just be secondary products. For example, photolysis of CH_3OH produces a small fraction of H_2CO. CO is produced from photolysis of CO_2, H_2CO and CH_3OH. Calculations show, however, that these processes cannot explain observed distributions. Direct production of molecules from grains has also been suggested, but here too there is a difficulty. Ice grains, especially those made from CO ice, are unlikely to survive out to distances of 10 000 km. At 1 AU, such grains would almost instantaneously sublime. Other processes are possible. For example, bombardment by UV photons or high energy ions from the Sun might tear molecules from the refractory organic mantle of cometary grains. Such processes would be relatively inefficient. Here then is one of the mysteries of cometary chemistry which remains unsolved.

Special attention should be paid to the HCN–HNC pair. HNC (hydrogen isocyanide) is a metastable isomeric form of HCN: in usual laboratory conditions, HNC converts rapidly into HCN. However, HNC is found in interstellar clouds with high abundances relative to HCN ([HNC]/[HCN] is typically 0.01 to 1). HNC was first discovered through its radio lines in comet Hyakutake with [HNC]/[HCN] = 0.06. In comet Hale–Bopp, unexpectedly, the ratio [HNC]/[HCN] was found to increase from 0.03 to about 0.20 as the comet approached the Sun (see Fig. 7.2). This variation cannot be understood by production from cometary ices. It has been argued that chemical reactions within the coma (and especially charge-exchange reactions with cometary ions) favoured by the very high density of this productive comet, could yield the formation of HNC from HCN. Indeed, calculations using a network of chemical reactions were able to account – at least qualitatively – for the observed HNC abundance and its variation in comet Hale–Bopp. In this case, the HNC abundance in the coma is not related to its abundance in cometary ices. Another remark in the same vein is that some of the minor species listed in Table 9.1 might not be primary species coming from nuclear ices: they might have been synthesized in the coma.

9.2 The diversity of comets

How much uniformity is there in the cometary population? Is it possible to define the typical composition of a comet? Or are there significantly different cometary compositions, expressing diversity in origins or history? These questions are clearly fundamental if we wish to understand the way comets formed, and yet it is still difficult to answer them.

There are still not many comets for which direct observations of parent molecules have been possible, so a statistical analysis cannot be made. Observed relative abundances [CO]/[H_2O] range from less than 2% (corresponding to absence of detection) to 20% for comet West, and even more for comets P/Schwassmann–Wachmann 1 and Hale–Bopp far from the Sun. In the case of P/Schwassmann–Wachmann 1, subsequent radio observations revealed CO, whilst H_2O and its dissociation products had not yet been detected. But we must be careful what we deduce about the actual relative abundance [CO]/[H_2O] in cometary ices. Firstly, we have not yet resolved the problem of the extended CO source mentioned in Chapters 4 and 5; and in addition, there is no reliable model for fractional sublimation of two molecules which sublime in such different conditions as CO and H_2O. Among the other parent molecules recently observed in the radio and infrared, the HCN abundance seems to vary between 0.03% and 0.2%. Variations in methanol abundance are greater and better established, with [CH_3OH]/[H_2O] ranging from 1% to 8% for about ten comets.

An indirect but more systematic study was carried out on spectrophotometric observations of OH, C_2, C_3, CN and NH radicals (see Sections 4.1 and 4.2). It concerned a uniform set of measurements on about a hundred comets. Given that OH comes essentially from H_2O, its abundance can be taken as an indicator for the abundance of water. Analysis shows that relative abundances [CN]/[OH] are more or less constant, grouped around a mean value of about 0.3%. In contrast, the ratio [C_2]/[OH] shows much greater spread, with variations up to a factor of almost a hundred observed from one comet to the next.

Extremes of behaviour have been observed in certain comets. Comet P/Wolf–Harrington 1991 V had practically no C_2 and C_3 bands, although it did have a strong CN band. Comet Yanaka 1988 XXIV had no detectable C_2 or CN bands, but had strong NH_2 bands (see Fig. 9.1). This implies a radically different ice composition in the two comets.

It would also seem possible that cometary nuclei are not chemically uniform. For example, volatile composition may differ from one active zone to another.

We are forced to conclude that the composition of cometary ices varies from one object to another. This makes it difficult to define a typical comet. What is the significance of this? Did comets form in different regions of the pre-solar nebula in which chemical composition was different? Or did they evolve in different ways, so that some became deficient in volatiles? We might also imagine that new comets, entering the inner Solar System for the first time, would be covered by a layer of ice whose surface has been chemically transformed as a result of secular UV and cosmic ray bombardment. In contrast, periodic comets would have undergone intense erosion of their surface ices. At each passage, deeper and deeper layers would be exposed. These layers would be better preserved and closer to their original chemical composition.

Efforts have been made to search for a possible correlation between the *type* of comet (new, old, Halley-type or Jupiter family) and their *composition*. However, no definite result has been established, except that Jupiter family comets seem to be carbon-depleted. This is shown by the lower value of their [C_2]/[OH] ratio. It is still unknown how this could stem from a difference in the actual nuclear ice composition.

More complete measurements will have to be made, particularly of parent molecules, and over a wider population of comets. *In situ* measurements by future space explorations will certainly clarify the question of nuclear

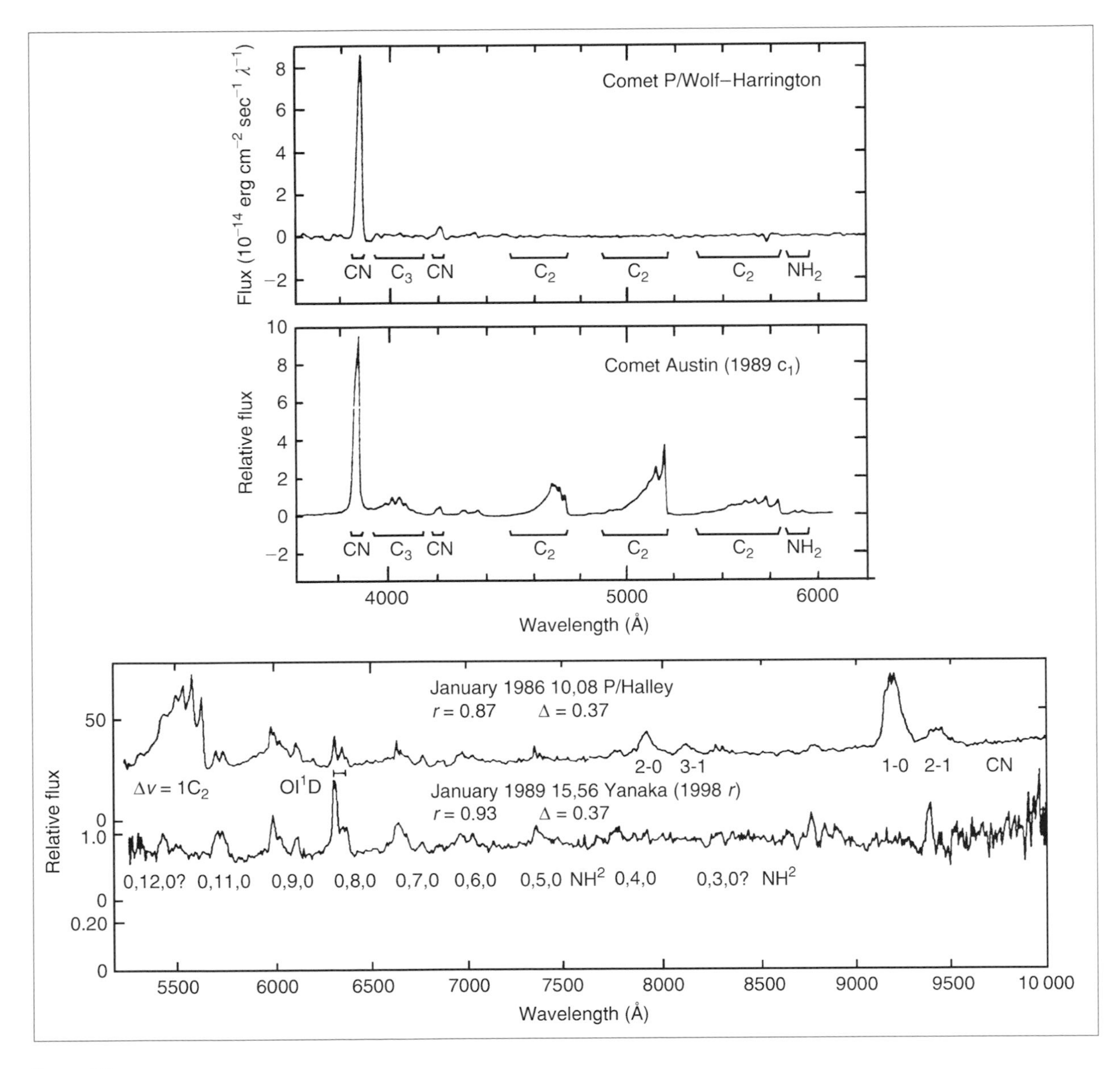

Figure 9.1.
Comparative spectra of 'normal' comets (P/Halley and Austin 1990 V) and 'pathological' comets such as P/Wolf–Harrington 1991 V, with weak C_2 band, and Yanaka 1988 XIV, with unusually intense NH_2 bands. From Fink, 1992 [45] and Schleicher *et al.*, 1993 [76].

composition. Unfortunately they will only concern a very limited number of objects, all of which will be short period comets (at least in the medium term). Systematic ground-based observation will be the only way to follow large populations of comets.

9.3 The analogy with interstellar molecules

Did comets form directly from the interstellar medium? In order to answer this question, we must first review what is known about the chemical composition of interstellar matter.

9.3.1 Interstellar gas molecules

Table 9.2 shows the interstellar molecules identified up to the present time. Among these, some have been observed in the visible spectra of stars. Absorption lines indicate the presence of interstellar molecular clouds between the stars and Earth. This is the case for the CH radical, which has been known since the 1940s. However, most molecules have been detected by their rotational lines in the radio region, mainly since the 1970s when millimetre radio astronomy came into being. The more recent development of telescopes equipped with high performance infrared spectrometers has made it possible to detect non-polar molecules like CH_4. These have no radio lines.

A glance at Table 9.2 will convince us that many radicals (highly reactive and unstable fragments of molecules) are present. Similarly for molecular ions (molecules from which an electron has been removed), which are also very reactive. Indeed, the interstellar medium is far from equilibrium. Physical conditions there are quite unlike those in a terrestrial laboratory. Stellar UV radiation tends to break molecules, except those deeply embedded in *dark nebulas*, where a high density of dust shelters them from destruction. Densities range from 1 atom to more than 10^6 atoms per cm^3. At such low densities, collisions are rare and chemical reactions can only reach equilibrium on a cosmic time scale.

There are many kinds of interstellar clouds. The least dense, containing only about 1 atom per cm^3, are the *diffuse clouds*. They contain almost no molecules, most of their matter being in atomic form. *Dense clouds*, containing about 10^6 atoms per cm^3 are essentially made up of molecular hydrogen, H_2, and other molecules. They can be very cold (only a few kelvins) but are sometimes violently heated when star formation begins. The first stage in star formation is gravitational contraction, which is one cause of heating. Then, once the stars have formed, they begin to produce their own heat from thermonuclear reactions. This heat is transferred to the surrounding interstellar medium (see Chapter 10).

The resulting very hot regions in the neighbourhood of young stars are called HII regions. Not only are molecules dissociated, but atoms are ionised by stellar radiation. Bright nebulas are produced in this way, such as part of the Orion nebula. At a certain stage in their evolution, some stars release matter at very high speed. This is the *stellar wind*, which also exists for the Sun, although the flux is relatively low. Shock waves occur where the stellar wind emitted by these stars comes into contact with the surrounding interstellar medium. Certain chemical reactions can take place in the shock wave which would be impossible in an unperturbed interstellar cloud. In particular, endothermic reactions requiring an energy threshold can occur. We can thus explain how certain molecules form in these *circumstellar regions*. Examples are SiO and some sulphur-bearing molecules.

9.3.2 Interstellar grains

Interstellar gas makes up about 10% of the mass of our Galaxy. Apart from this, a significant part of interstellar matter exists in the form of dust particles, or *interstellar grains*. This has been known for a long time, because dust has the property of absorbing light from the stars. Dark clouds, either diffuse or in the form of compact blobs (Bok globules), appear in images of star fields within the Galaxy.

Interstellar grains also manifest their presence by a reddening of the light from stars. Because they are small, they tend to absorb radiation at short wavelengths (blue light) better than that

Table 9.2. List of interstellar molecules.

Inorganic molecules							
Diatomic							
H_2	SiO	HCl	**CO**	SiN	NaCl	**CS**	SiS
AlCl	NO	O_2?	KCl	**NS**	PN	AlF	HF
Triatomic							
H_2O	**H_2S**	**CO_2**	NO_2	**SO_2**	**OCS**	NaCN	
4 atoms							
NH_3							
5 atoms							
SiH_4							

Organic molecules			
Acids		Alcohols	
HCN	Hydrogen cyanide	**CH_3OH**	Methanol
HNCO	Isocyanic acid	C_2H_5OH	Ethanol
HCOOH	Formic acid		
CH_3COOH	Acetic acid		
Aldehydes and ketones		Hydrocarbons	
H_2CO	Formaldehyde	**CH_4**	Methane
H_2CCO	Ketene	**C_2H_2**	Acetylene
CH_3CHO	Acetaldehyde	C_2H_4	Ethylene
$(CH_3)_2CO$?	Acetone	CH_3CCH	Propyne
		CH_3C_4H	Methyl diacetylene
Amides		Esters and ethers	
HN_2CHO	Formamide	**$HCOOCH_3$**	Methyl formate
NH_2CH_3	Methylamine	$(CH_3)_2O$	Dimethyl ether
NH_2CN	Cyanamide	C_2H_5O	Ethylene oxide
Organo-sulphurous molecules		Other cyanides	
H_2CS	Thioformaldehyde	**HC_3N**	Cyanoacetylene
HNCS	Isothiocyanic acid	**CH_3CN**	Methyl cyanide
CH_3SH	Methyl mercaptan	C_2H_5CN	Ethyl cyanide
		CH_3C_3N	Methyl cyanoacetylene
		CH_2CHCN	Vinyl cyanide
		HC_3HO	Propynal

Unstable molecules							
Radicals							
CH	**CN**	**OH**	**SO**	HCO	**NH**	**NH_2**	HNO
MgNC	MgCN	CP	SiC	CH_2	CH_2CN	CH_2N	
Ions							
CH^+	**CO^+**	SO^+	N_2H^+	H_3^+	**H_3O^+**	$HCNH^+$	HC_3NH^+
HCO^+	HOC^+	H_2COH^+	$HOCO^+$	HCS^+	$HOCS^+$?		
Carbon chain molecules							
C_2	**C_3**	C_5	C_2H	C_3H	C_4H	C_5	C_6H
C_7H	C_8H	C_3H_2	C_4H_2	C_6H_2	C_2O	C_3O	C_2S
C_3S	C_5S?	C_4Si	C_3N	C_5N	HC_5N	HC_7N	HC_9N
$HC_{11}N$							
Cyclic molecules							
c-SiC_2	c-C_3H	c-C_3H_2					
Isomers							
HNC	CH_3NC?	HCCNC	HNCCC				

Bold face indicates detection in comets. The molecular ions OH^+, H_2O^+, N_2^+ and CO_2^+, observed in the visible spectra of comets, have not yet been detected in the interstellar medium.

at longer wavelengths (red light). They also cause a polarisation phenomenon in the light from stars. We must assume that these grains have an elongated shape and are oriented by interstellar magnetic fields. Hence they are able to selectively absorb certain polarisations of light. These grains have sizes between 10^{-3} μm and 0.5 μm. It is quite possible that much bigger grains exist (almost planetesimals), but which have escaped observation up to the present.

Interstellar grains are sometimes observed very close to stars, where the temperature is relatively high. We can deduce that they are composed of *refractory molecules*, that is, mineral molecules which are particularly resistant to temperature. It is thought that refractory grains are ejected from stars during certain advanced stages of their evolution, i.e. when thermonuclear reactions have produced heavy elements by nucleosynthesis. Refractory molecules differ according to the type of star which produced them. Hence, old giant stars, rich in oxygen, are a source of silicate grains such as iron and magnesium silicates. Fe_2SiO_4 and Mg_2SiO_4, mix together to form the terrestrial mineral known as olivine, whereas $FeSiO_3$ and $MgSiO_3$ blend to form pyroxene. Carbon-containing stars give grains of amorphous carbon. Novas and supernovas eject a significant proportion of their mass during an explosion (a rather extreme form of solar wind!). In particular, they eject iron- and metal-rich grains (e.g. containing FeS, FeC, MgS, Al_2O_3). These refractory minerals can be detected by their spectra. Silicate absorption bands at 9.7 and 18 μm lie in the infrared, and a feature in the UV at 0.217 μm is due to graphite.

When these grains arrive in cold, dense molecular clouds, they are covered by interstellar molecules which condense on their surface to form an icy mantle. They thereby become areas of intense interstellar chemistry. Although stable molecules like H_2O, CO and CO_2 can survive intact on grains, the highly reactive radicals and ions will contribute to synthesis of more complex molecules. In this case, grains act as catalysts. Some molecules may then be released into the interstellar medium. It is thought that molecular hydrogen, H_2, is produced in this way from hydrogen atoms in the interstellar gas. It is extremely abundant in molecular clouds. In addition, ice molecules are subjected to interstellar UV radiation and bombarded by cosmic rays over secular time scales. These molecules can also evolve. Organic molecules polymerise and form refractory or semi-refractory mantles. Such processes have been simulated in the laboratory (see Chapter 7).

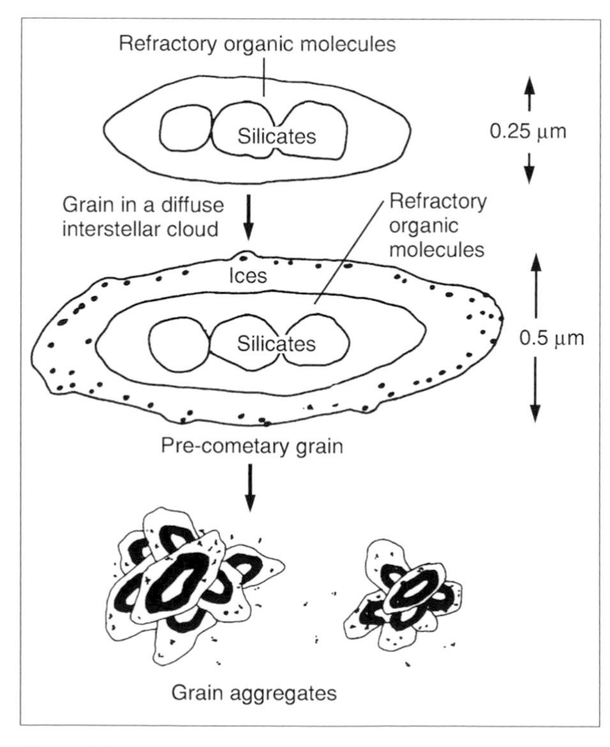

Figure 9.2.
From interstellar dust grains to cometary grains, according to the model of J.M. Greenberg. Greenberg, 1990 [47].

Radio spectroscopy is of little help in deciphering the composition of icy grains. This is because rotational transitions do not occur in the solid phase. The question has to be dealt with by infrared spectroscopy. In addition to previous ground-based investigations, decisive results were obtained with the ISO satellite (see Fig. 9.3). Our knowledge in this area is listed in Table 9.3. Many uncertainties remain. The infrared spectra of ices are more difficult to diagnose than radio spectra of gases. Spectral signatures are wide, often mixed up together and affected by mutual influences between ice compounds. This means that the spectrum of an ice mixture is not just the sum of the spectra of its individual components, as would happen in a mixture of gases.

Some compounds have not yet been identified with certainty, in particular H_2S, which is the subject of some controversy. Certain observed features remain unidentified, although they can sometimes be attributed to a particular functional group, e.g. X—CN or X—CH, without specifying exactly which molecules are involved. Abundance assessments are still rather speculative, even contradictory between different studies. Improvements will come partly through more sensitive observations, partly

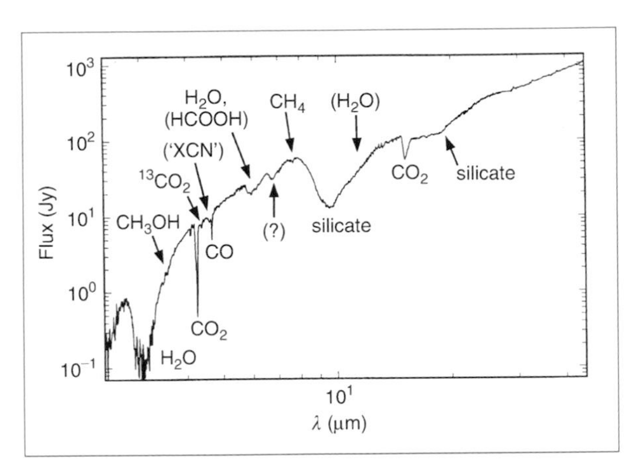

Figure 9.3.
The infrared spectrum of IRS9, a deeply embedded protostar, observed by ISO, showing in absorption the features of interstellar ices and silicates. From Whittet *et al.*, 1996 [81].

Table 9.3. Relative abundances of ices in the interstellar medium and in comets.

Species	Interstellar ices	Cometary volatiles
H_2O	= 100	= 100
CO	10–40	2–20
CH_3OH	5	1–7
CO_2	10	2–6
H_2CO	2–6 ?	~ 1
HCOOH	3 ?	~ 0.1
CH_4	1–2	~ 1
Other hydrocarbons	?	~ 1 C_2H_2, C_2H_6
NH_3	< 10	0.5
O_3	≤ 2	?
X–CN	≤ 0.5–10	0.37 Nitriles + HNCO
OCS, XCS	0.2	0.4 OCS + CS
SO_2	?	~ 0.1
H_2	> ~ 1	?
N_2	?	?
O_2	?	?

through laboratory study of infrared spectra of compounds analogous to those in interstellar grains, which could serve as a model.

9.3.3 Aromatics

Another form of interstellar matter was discovered at the beginning of the 1980s. It was revealed through a series of emission bands at 3.3, 6.2, 7.7, 8.7 and 11.3 μm, which are characteristic of the CH functional group in unsaturated carbon chain molecules. These signals can be attributed either to large molecules or to very small grains. This implies, therefore, an interstellar component lying somewhere between classic gas molecules and dust grains. The substance absorbs ultraviolet photons without dissociating. It re-emits their energy with remarkable efficiency in the form of infrared photons corresponding to vibrational modes of the molecule.

More specifically, several types of molecule have been put forward to explain this component:

* *polyaromatic hydrocarbons* (PAHs) are molecules containing several benzene rings (benzene C_6H_6 is the archetypal cyclic aromatic);
* hydrogenated amorphous carbon, consisting of graphite fragments with hydrogen atoms bonded onto the surface;
* coal dust particles, similar to those in terrestrial coal mines;
* a whole class of organic residues, similar in the terrestrial context to heavy tars and soot particles.

The distinction between these substances is sometimes subtle and may well be the result of rather academic debate. The interstellar medium is very probably a poorly defined mixture of complex substances which will be difficult to identify in detail. The important thing is to recognise their common properties: high unsaturated carbon content; great stability under UV radiation; specific infrared spectrum; and lastly, carcinogenic effects on animals and human beings!

Let us note that this component occurs in comets (see Chapter 6). Certain PAHs have also been identified in the organic matter of carbonaceous chondrites, the most primitive meteorites.

9.3.4 Isotopic ratios

The deuterium to hydrogen isotopic abundance ratio ([D]/[H]) has been measured in the water of comet Halley by mass spectroscopy (Section 5.4.3) and by radio spectroscopy in comets Hyakutake (Fig. 6.12) and Hale–Bopp. The three values are remarkably similar: 0.0003, which is a factor of ten higher than the value for the protosolar nebula, and about twice that for water in Earth's oceans (Fig. 9.4). Similar, or even higher enhancements of the [D]/[H] ratio have also been observed in interstellar water and other interstellar molecules such as HCN, H_2CO, and CH_3OH.

In contrast, the isotopic abundances for heavier atoms – C, N, O, S – are found 'normal' compared to cosmic values. Isotopic enhancements in molecules are due to fractionation effects in chemical reactions, mainly reactions between ions and neutrals that occured at the low temperatures encountered in the interstellar medium. Such fractionation effects are much more important for the deuterium–hydrogen pair, which has a large mass difference, than for isotopes of heavier atoms (such as ^{12}C and ^{13}C) which have comparable masses.

The high [D]/[H] ratio in cometary water suggests that cometary material comes from interstellar matter which was not at all – or only slightly – processed in the presolar nebula. In

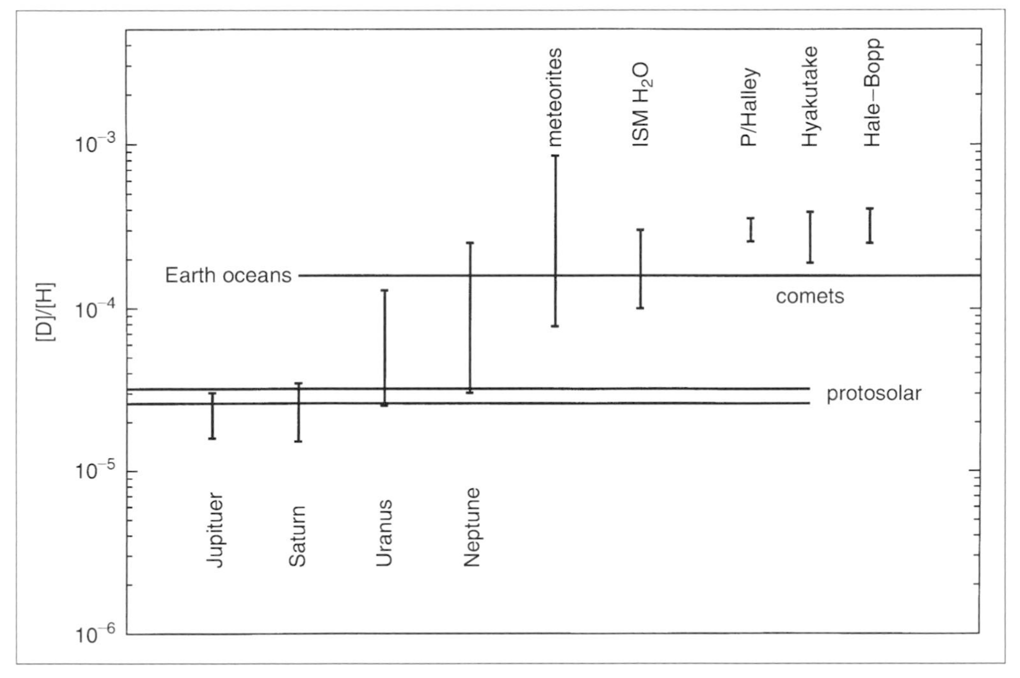

Figure 9.4.
Measurements of the deuterium to hydrogen isotopic abundance ratio for several objects in the outer Solar System and for meteorites. The terrestrial value (measured in the oceans) is given for comparison, as are those for the protosolar nebula and the local interstellar medium. A progressive deuterium enrichment is observed in moving outwards from the nearer giant planets. The D/H ratio is also notably higher in meteorites and comets. The generally accepted explanation is that deuterium trapped in ices may have been the result of enrichment processes involving ion–molecule reactions in the interstellar medium, before bodies formed in the Solar System. It is this deuterium which is observed on Titan, on comet Halley and in meteorites, the relics of asteroids or comets. On the other hand, deuterium observed on the giant planets would come partly from the HD molecule and partly from ices trapped in their nuclei. In the case of Jupiter and Saturn, contributions from the nuclei would be low, hence a value for the D/H ratio compatible with that of the protosolar nebula. But in the case of Uranus and Neptune, contributions from the nuclei would be more significant, hence a deuterium enrichment relative to values for the protosolar nebula. (Adapted from Bockelée-Morvan *et al.*, 1998 [32].)

particular, exchanges with molecular hydrogen, which could have re-equilibrated the [D]/[H] ratio, would not have been significant.

9.3.5 Comparing interstellar and cometary matter

There is a remarkable similarity of composition between cometary ices and icy mantles of interstellar grains (compare Tables 9.1 and 9.3). Not only do they contain the same compounds (H_2O, CO, CO_2, CH_3OH, CH_4, etc.), but their relative abundances are also comparable. This is true whatever means of observation is used: solid state absorption features in infrared spectroscopy of interstellar grains; infrared fluorescence and radio rotational spectra in the gas phase for cometary molecules. Moreover, refractory cores of interstellar and cometary grains are also very similar. They both contain molecules of silicates and high abundances of metallic elements.

It is natural to conclude that we are dealing with the same kind of material and that cometary nuclei result from accumulation of practically unmodified interstellar grains. This hypothesis has been supported by J.M. Greenberg and his team at the laboratory astrophysics group of the Leyden Observatory for more than 20 years now. In reality, the situation may not be so simple. Other hypotheses have been put forward, as we shall see in the next chapter concerning the origins of the Solar System. Grains similar to those in the interstellar medium may have formed in the primitive solar nebula from matter which had previously undergone chemical reactions and thereby lost its primitive composition. It is also possible that interstellar grains in the pre-solar nebula could have lost their original icy mantles through heating. Then, in a second stage, mantles might have recondensed. In this case, only the refractory nuclei would be 'authentic'.

9.3.6 Condensation of ices

Let us accept the idea that the present constituents of cometary ices come mainly from condensation of molecules in the gas phase, rather than chemical reactions on grain surfaces, the very nature of these constituents can tell us about conditions prevailing during the formation of cometary grains. In particular, we can learn something about temperature conditions. Indeed, we find molecules like water which are not very volatile, together with some rather volatile molecules like H_2S or CO; but highly volatile molecules such as N_2 and H_2 would not appear to be present. Table 9.4 gives sublimation/condensation temperatures for some molecules occurring in comets. These temperatures have been calculated for extremely low densities (10^{13} molecules/cm^3, almost a perfect vacuum), such as would prevail in the pre-solar nebula. These conditions are quite unlike those found on Earth. Hence water condenses at 150 K, well below the value of 273 K we are used to on Earth! From the table, we see that cometary H_2S could not have condensed above 55 K, CH_4 could not have condensed above 31 K and CO could not have condensed above 25 K. This gives some idea of the extreme temperature conditions operating when cometary grains were forming.

Temperatures in Table 9.4 are for equilibrium between the relevant ice and its own vapour. Values can be quite different for a mixture of ices. We then observe the phenomenon of fractional sublimation. Each substance can sublime

Table 9.4. Sublimation equilibrium temperatures for some molecules and noble gases.

	Sublimation temperature (K)
Molecules	
H_2O	152
CH_3OH	99
HCN	95
SO_2	83
NH_3	78
CO_2	72
H_2CO	64
H_2S	57
CH_4	31
CO	24
N_2	22
H_2	5
Noble gases	
Ar	(87)
Ne	(27)
He	(4)

We assume a total density 10^{13} cm^{-3}, mainly due to H_2 molecules, and relative abundances ranging from 10^{-4} for water to 10^{-8} for less abundant molecules. These values correspond to conditions expected in the pre-solar nebula. However, even a change of several orders of magnitude would hardly affect the temperatures.

separately, but attractive forces between molecules play an important role. At short range, this attraction may be due to van der Waals forces, or electrostatic forces in the case of polar molecules. These are the same forces that govern surface tension in liquids, cohesion in crystals and the dissolving of one substance in another. Highly polar molecules like NH_3 and H_2S have strong affinity for water (an appreciable quantity can be dissolved in liquid water). Although these molecules are highly volatile, we would expect water to retain them. We would therefore not expect them to escape from cometary ices unless the water itself sublimed. This is not so for the weakly polar CO molecule, which can migrate through water ice and sublime as soon as the temperature rises above 25 K. However, laboratory experiments have shown that a significant fraction (several per cent) of CO can remain trapped in water ice, even up to a temperature of 150 K. This may explain why we find carbon monoxide in the nuclei of periodic comets, even though the inner nucleus has a temperature well above 25 K.

By the same reasoning, it seems unlikely that, if cometary nuclei formed at temperatures above a few kelvins, they could have retained highly volatile non-polar constituents such as molecular hydrogen, H_2, or the lighter noble gases (helium, neon, argon). This could also explain the very low abundance of molecular nitrogen, N_2.

There is another way that molecules might be trapped in water ice, namely natural cages formed by its crystals. One molecule of CO, N_2 or CH_4 could be held in such a cage for every five or ten water molecules. This forms the so-called *clathrate hydrates*, which have been well studied experimentally. Their possible relevance was first proposed in the 1950s by Fred Whipple and Armand Delsemme. They hoped to explain the simultaneous release from cometary nuclei of H_2O and other parent molecules. At the present time, the hypothesis of cometary clathrates has encountered several difficulties. It would seem that conditions in which cometary ices formed are not compatible with the stability of these clathrates. In addition, the proportion of host sites available in ice crystals is not sufficient to explain the high observed abundances of molecules like CO, CO_2 and CH_3OH.

In conclusion, we can sum up the main points:

* cometary ices are very close in composition to icy mantles of interstellar grains;
* the nature of cometary ices implies that they formed at very low temperatures;
* fractionation of cometary volatiles occurred in several stages. At condensation, the most volatile gases (H_2, He, Ne and maybe N_2) would not have been retained. During secular evolution of grains and cometary nuclei, the more mobile and volatile molecules (N_2 and possibly CO) were depleted. During cometary activity, volatiles were not necessarily all released simultaneously.

We must be careful not to assume that the composition observed in the coma, often at heliocentric distances close to 1 AU, is the same as that of ices in the nucleus; and we should not therefore draw too hasty conclusions about primordial abundances.

10
Comets and the history of the Solar System

10.1 Asteroids and their relation to comets

Just a few years ago, the known Solar System contained nine planets accompanied by a handful of satellites, several thousand asteroids almost all concentrated between Mars and Jupiter, and the irascible comets. As a result of progress in ground-based observation and space exploration, we have found a much richer and more varied Solar System. In addition, the simplistic classification has evolved. Dividing lines between the various classes of object have become less clear.

The larger satellites (the Moon, Galilean satellites, Saturn's Titan) are comparable to or bigger than the smallest planets, Mercury and Pluto. Asteroids catalogued so far have sizes ranging over three orders of magnitude. It is clear that our knowledge of the smaller asteroids is limited by observational capacity. The largest asteroids are similar in size to medium-sized satellites and it has been more or less established that many small satellites are in fact captured asteroids. This must be the case for Phobos and Deimos, the two satellites of Mars, as well as Phoebe, one of Saturn's moons, which has the particularity of orbiting in the opposite direction to all the others. Several satellites of Jupiter, Saturn and the other outer planets are covered with water ice. One indicator for this is their high reflectivity or albedo, reaching values of 0.5 or more. (The record is held by Enceladus, a satellite of Saturn, with albedo 1.) Another indicator is infrared spectral signature. Their densities, when they are known, are low. They lie in the range 1.2–1.5, suggesting that their interiors are also mainly composed of ice. Ganymede and Callisto, the giant satellites of Jupiter, are also covered with ices. However, they have higher densities, in the range 1.8–1.9. This suggests a mixture of ices and minerals (silicates). These bodies must therefore have formed from the same kind of ingredients as we find in comets. They differ profoundly, however, in the metamorphic transformations they have undergone since formation. They have been heated, first by gravitational energy during accretion, then through radioactivity of elements in their nuclei, and maybe also by internal friction caused by tidal forces. Such bodies thus have a thermal history. Their ices may have melted, at least partially, and it may be that a liquid phase still exists inside some of them. Large gravitational effects, which have led to their perfectly spherical shape, may have been sufficient to cause segregation of the various constituents through sedimentation.

In order to make an exhaustive inventory of the smaller objects in the Solar System, we should also mention planetary rings. These are

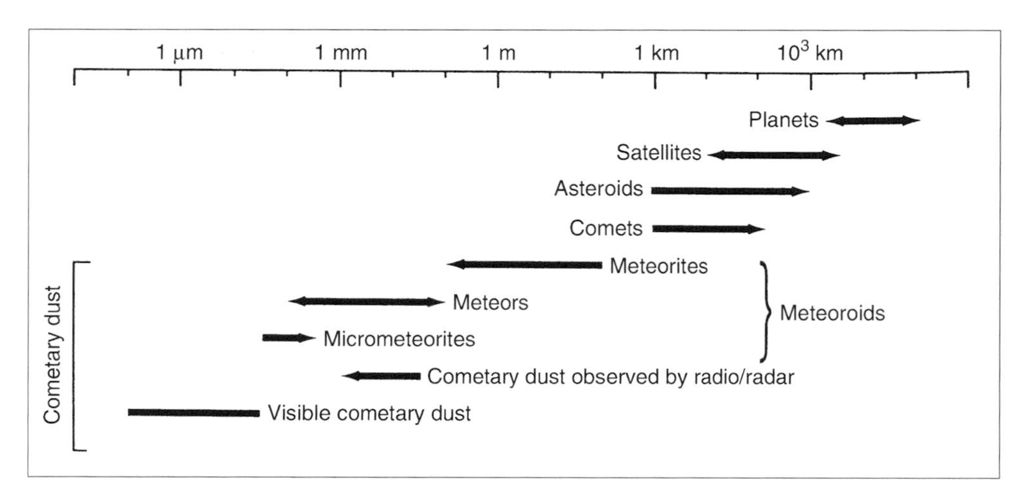

Figure 10.1.
Sizes of objects in the Solar System.

observed around all the giant planets, Jupiter, Saturn, Uranus and Neptune. They are almost certainly supplied with dust from comets and asteroids.

Asteroids, or minor planets, were first discovered during systematic searches for new planets undertaken at the end of the eighteenth century. The first asteroid, Ceres, was found on 1 January 1801 by Guiseppe Piazzi (1746–1826) of Palermo. Five asteroids were known in 1845, then 300 in 1890. The rate of discovery increased enormously from this time with the development of photographic techniques and systematic search programmes.

Some asteroids seem to be rather similar to cometary nuclei. These are the D type asteroids (see the Box entitled *Nomenclature and classification of asteroids*). They are characterised by their low albedos, in the range 0.02–0.06. These are among the lowest observed for asteroids, but quite comparable with those of cometary nuclei. The slope of their spectra reveals a red colour in the visible, but it becomes flat in the near infrared. This may indicate a layer of refractory organic residues on the surface. Such a layer could come from effects of solar ultraviolet radiation and cosmic rays on more primordial molecules. A terrestrial model of such a substance is kerogen, a sort of natural tar found in bituminous shales and which exhibits an analogous spectrum.

Diameters of known asteroids range continuously from a maximum of 900 km for Ceres, to less than 1 km. It is quite clear that many small objects still escape today's observational capabilities. Asteroids are too small to retain any atmosphere.

A spacecraft flyby is the best way to find out about an asteroid, directly observing its shape and rotational state. Certain missions are likely to come close to one or more asteroids on the way to their primary target (see Table A.3, *Space missions*, in Appendix). On its way to Jupiter, Galileo flew by Gaspra and Ida (see Figs. 10.3, 10.4). Pictures of Ida revealed that this asteroid has a companion, later named Dactyl, gravitating around it. It is only 1.5 km in diameter. We do not know what fraction of asteroids belong to these double systems nor how they came together. These mysteries remain for future missions to resolve.

Nomenclature and classification of asteroids

When discovered, asteroids (also known as minor planets) are given a provisional denomination containing the year of discovery followed by two capital letters. The first letter indicates the half-month of discovery, and the second, the order of discovery within this period. With present techniques, more than 10 000 new asteroids are discovered each year. However, the orbit can be determined only for a very few of them. When an orbit has been established to within certain confidence levels, the asteroid is attributed a permanent number. At the present time, more than 9000 asteroids have been numbered in this way. A name can then be suggested by the discoverer (or else attributed by a special commission of the International Astronomical Union). More than 6000 asteroids named in this way demonstrate the fertile imagination of those involved: mythological names, famous people, discoverer's family and friends, geographical names, and many others. To give an example, asteroid 1977 UB was the second asteroid discovered in the first fortnight of November 1977. Its discoverer was Charles Kowal and it was eventually numbered 2060 and called Chiron.

Asteroids are classified in two ways: according to orbit or according to spectral features (taxonomic classification).

Classification by orbit

Figure 10.2 shows the distribution of semi-major axes of asteroidal orbits. The great majority of them are located in the Main Belt, between Mars and Jupiter, hence the broken planet hypothesis originally put forward. Some asteroids, known as the Trojan asteroids, are actually located on the same orbit as Jupiter, near the Lagrange points. These are points 60° before or after the planet on its orbit, where celestial mechanics predicts a stable orbit for small objects. We shall discuss more distant asteroids, the Centaurs, shortly (see Section 10.1.2). For the moment, only a very small number have been found.

Closer to Earth, we find the asteroids of the inner Solar System. Some have orbits which come within the orbit of the Earth and are known as Earth-crossing asteroids, or ECA. Examples are the Apollo and Amor families, named after their typical representatives. Also close is the Aten family, between Mars and Earth. These asteroids sometimes come very near to Earth and the smaller members of the family (less than 1 km across) are then observable. They are the subject of very careful observation, through fear of a possible collision with Earth.

The distribution of the various families is largely governed by gravitational effects of planets. In particular, the Main Belt was to a large extent shaped by Jupiter. A resonance phenomenon dictates that orbital periods commensurable with Jupiter's are forbidden. The result is a series of gaps in the distribution of asteroidal semi-major axes (see Fig. 10.2). These are known as the Kirkwood gaps. Among several groups set apart in this way, we could mention the Hilda family, all outside the Main Belt, and the Hungarians which all lie within it.

Taxonomic types

Classification in terms of taxonomic types is based on spectra and is therefore intended to reflect surface mineralogical composition. In this way, we can compare the spectra of asteroids with those of analogous terrestrial rocks or meteorites. This classification consists in attributing a given letter to each type identified. Without going into the details, note that we can distinguish *evolved or igneous* asteroids, which have resisted relatively high temperatures. These are mainly the stony or S type and the metallic or M type. They occur principally in the inner regions of the Main Belt and in the Apollo–Aten–Amor families, i.e. they are the ones nearest to the Sun. Then there are the *primitive* asteroids which populate most of the outer region of the Main Belt and the Trojans: type C, the carbonaceous asteroids, which may be related to the carbonaceous chondrite meteorites, and type D.

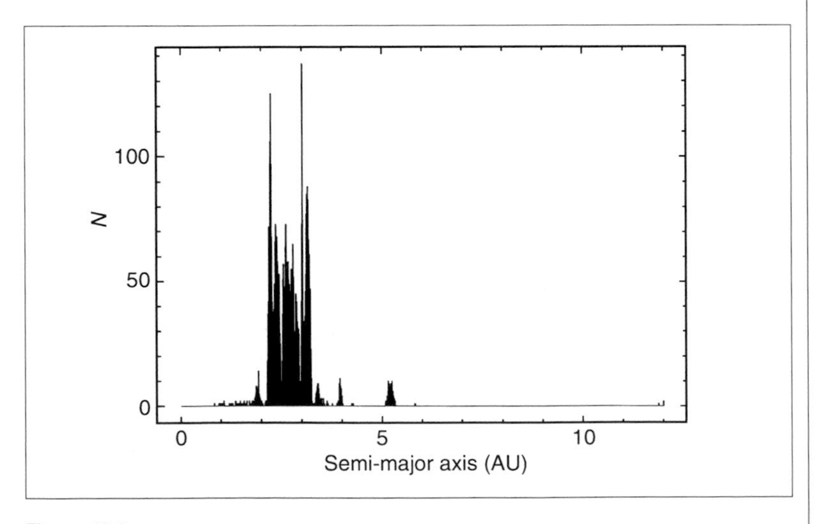

Figure 10.2.
Distribution of asteroidal semi-major axes.

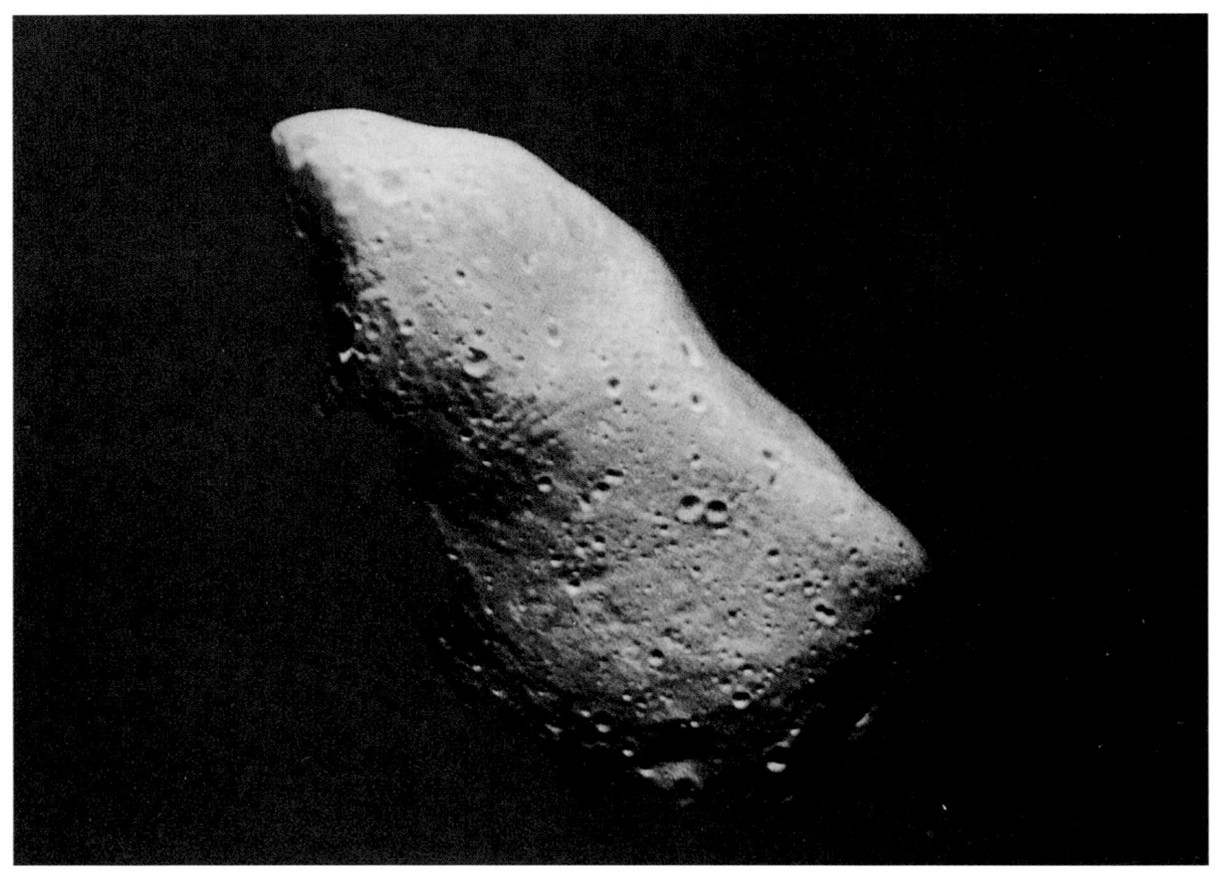

Figure 10.3.
Photo of the asteroid Gaspra taken by the Galileo probe in October 1991 at a distance of 5300 km. It measures about 19 × 11 km. Colours have been exaggerated by image processing to bring out mineralogical variations across the surface. Document courtesy of NASA and the Jet Propulsion Laboratory, Caltech.

The Rosetta cometary mission will also fly close to asteroids (see Section 11.3); and the NEAR space mission (Near Earth Asteroid Rendezvous) aims to put a probe into orbit around Eros, a small Earth-crossing asteroid. On its way, it has already encountered Mathilde, a quite large Main Belt asteroid. Mathilde's gravitational field slightly altered the trajectory of the probe, leading to a measurement of the asteroid's mass and density. The value obtained, only 1.3, is astonishingly low, suggesting that this body is porous or that it contains ice. Both these possibilities are reminiscent of cometary nuclei.

10.1.1 Comet–asteroid relations

The distinction between comets and asteroids is not always very clear. At first it was thought that the two classes of objects had different origins and compositions: comets contained a large proportion of ices, and asteroids were mainly composed of refractory, stony and metallic minerals. However, such a distinction is not always easy to support with observation. Although active comets can be classed without further discussion, the same is not true for *dormant* or *dead comets*.

Figure 10.4.
Photo of Ida with its satellite Dactyl, taken by the Galileo probe during flyby at 10 900 km. Ida is about 56 km long and its satellite has diameter 1.5 km. Document courtesy of NASA and the Jet Propulsion Laboratory, Caltech.

Dormant comets are genuine cometary nuclei which are not at present active, either because they are too far from the Sun, or because active regions have been temporarily smothered by a dust deposit, protecting them from solar heating. Such a comet may become active again, possibly as a result of some modification to its orbit.

Dead comets have exhausted most of their volatile elements so that only a mass of refractory matter remains. Certain comets disappear between one passage and another, although their orbits are perfectly well determined. They can be supposed to have become dead comets.

Among the many known asteroids, there is of course an unknown fraction of dormant and dead comets. As a result, classification errors abound. Apart from Chiron, which we shall discuss later, we could mention P/Parker–Hartley 1987 XXXVI, identified with asteroid 1986 TF, and Wilson–Harrington 1949 III, which had to be withdrawn from the comet catalogue when it was identified as asteroid (4015) 1979 VA.

Orbits of certain asteroids strongly resemble those of short period comets. Thus, (944) Hidalgo has period 13 years and its orbit, with eccentricity 0.658, crosses Jupiter's. Several asteroids in the Apollo–Amor family, whose orbits intersect Earth's orbit, are associated with showers of shooting stars: (2101) Adonis, (2201) Oljato, and (3200) Phaeton. These asteroids must therefore have shed dust at some relatively recent stage. Could it have been during a phase of cometary activity? The search for a coma around the nucleus of these asteroids has proved fruitless.

10.1.2 Chiron and the Centaurs

Only a small number of asteroids are known beyond the Main Belt. Some have highly eccentric orbits, similar to cometary orbits, taking them temporarily beyond the orbit of Jupiter. This is the case for Hidalgo, for example. We may wonder whether these are not dead comets. There are very few asteroids gravitating between Jupiter and Neptune: at the present time, we know only of Chiron, Damocles, Pholus, Nessus and a few others discovered recently (1994 TA, 1995 DW_2, 1995 GO, 1997 CU_{26}). These are Centaurs, the name given to trans-Saturnian objects.

Chiron is a unique object. When it was discovered in 1977, nothing remarkable was noticed apart from its impressive dimensions. However, continued photometric observations later revealed an increase in brightness which could not be explained merely by the object's being nearer to the Sun. (Chiron gravitates at an average distance of 13.7 AU from the Sun, and reached its perihelion at only 8 AU in 1996.) Pictures show a faint coma and even the beginnings of a tail (see Fig. 10.5). It had to be admitted that Chiron, originally classified as an asteroid, was in fact a comet. Its activity seems to be sporadic, with successive bursts of activity

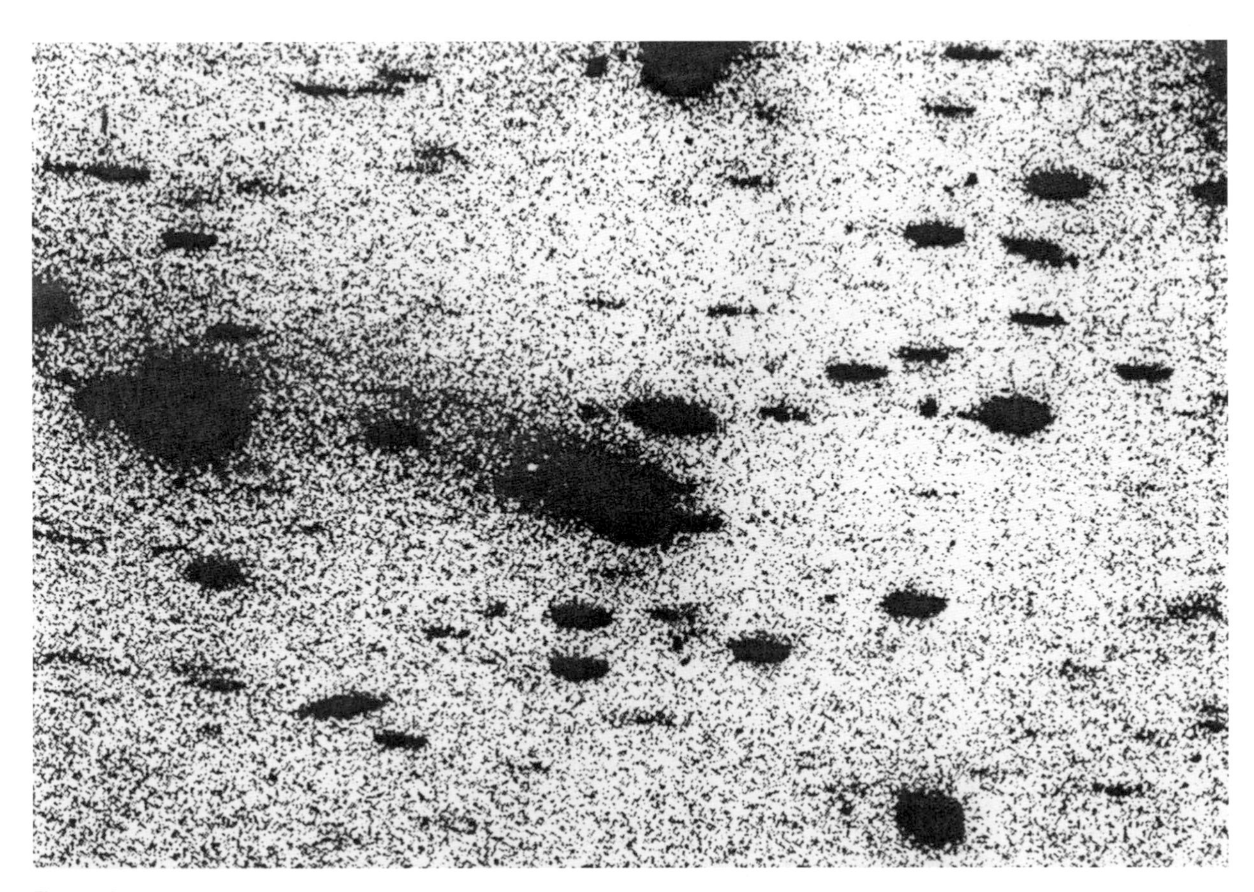

Figure 10.5.
Chiron during an active period. Photo taken with ESO's 3.4 m NTT on 25 December 1992. At the centre is Chiron's short cometary tail. From Hainaut, 1994 [49].

followed by calm periods. Mechanisms are probably similar to those of distant comets (see Chapter 7), governed by sublimation of highly volatile molecules such as CO. Photometric observations and stellar occultations established a diameter of 200 km. This is a very large comet. The gravity of its nucleus is far from negligible. Although it may not be sufficient to retain gases, it can quite likely hold the larger dust grains, thereby forming a captive dust atmosphere. Spectroscopic study of Chiron has revealed nothing about the nature of gases in its coma, apart from the presence of the CN radical. The latter was observed once during a period of maximum activity, but never again recorded.

Pholus, the second well studied Centaur, has never shown cometary activity since its discovery in 1992. It is very red in colour. Apart from Mars, it is the reddest object in the Solar System. The colour may be due to a layer of organic residues covering its surface, like the tholins. These result from high energy irradiation of CH_4 ice and ices of other organic molecules, for example by UV or interstellar cosmic rays. Such a semi-refractory layer may have protected inner volatile ices from heating, and prevented cometary activity, at least temporarily.

10.1.3 Trans-Neptunian objects

We still know very little about the outer Solar System. At 30 AU from the Sun, Neptune is the last planet of any reasonable size. Pluto, of diameter 2300 km, is comparable with the larger satellites. Its orbit has semi-major axis 39.5 AU but is very eccentric and sometimes comes within the orbit of Neptune (as is the case at the moment). Until 1992, these were the limits of the known Solar System.

The search for new planets has always been an obsession for astronomers. It led to the discovery of Uranus, at the end of the eighteenth century, by William Herschel (1738–1832) in England; then to the first minor planets at the beginning of the nineteenth century, when astronomers were searching for a hypothetical planet between Mars and Jupiter. Neptune was discovered by Johann Galle (1812–1910) at the Berlin observatory, using calculations made by Urbain Le Verrier (1811–1877) at the Paris observatory. And finally, Pluto was discovered by Clyde Tombaugh (1906–1997) in 1930, at the Flagstaff observatory (Arizona).

Does the Solar System suddenly come to an end at 40 AU? It certainly seems unlikely. As we shall see later, during formation of the Solar System, the protoplanetary accretion disk must have extended further. Even if it did not produce large planets, planetesimals formed from it must persist. Moreover, around 1950, Edgeworth and Kuiper put forward a hypothesis concerning short period comets. Most have orbits with low inclination and could not have originated in the spherically symmetric Oort cloud. It seemed more likely that they had come from a flat disk. This in turn gave birth to the idea of a ring of planetesimals and comets located beyond the orbit of Neptune: the Edgeworth–Kuiper belt.

Of course, the only way to check the hypothesis was to detect these objects. However, such distant bodies, completely devoid of any cometary activity, are not observable with today's observational means unless they measure at least 100 km across. The hunt for trans-Neptunian objects was nevertheless open. It required telescopes of diameter at least 2 m, equipped with CCD cameras of field as wide as possible. It also required a lot of observing time, stubbornness and faith.

The first discoveries were made by David Jewitt and Jane Luu following systematic exploration of almost one square degree of sky. They found the object 1992 QB_1 in August 1992 and then 1993 FW in March 1993. Subsequently, the rate of discoveries began to increase. Four objects were found in 1993 and ten in 1994. By the end of 1997, no fewer than sixty objects had been recorded. The majority of these were discovered by Jewitt and Luu's group, mainly with the Mauna Kea telescopes in Hawaii. Most of the others were discovered by a British team at La Palma observatory in the Canary Islands.

These objects are found by taking series of long exposures close to the ecliptic in the neighbourhood of the antisolar direction. Trans-Neptunian objects differ from Main Belt asteroids by their much slower displacement relative to the stars. Their rate of displacement can be used to evaluate their distance from the Sun. As a first approximation, orbits are assumed circular. Distances range from 35 AU (barely further than Neptune) to a little more than 45 AU (see Figs. 10.6 and 10.7) for objects found so far. In order to specify orbits more precisely, continued astrometric observations must be made available. Unfortunately, this task is as difficult and time-consuming as discovering the objects in the first place. Although many orbits have now been determined, other objects must be considered as lost.

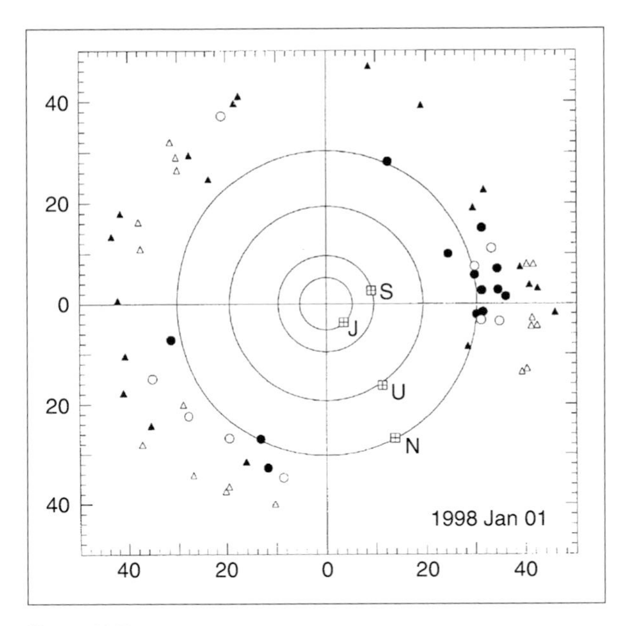

Figure 10.7.
Positions of the known trans-Neptunian objects as of 1 January 1998. 'Plutinos' are shown as *circles* and other objects as *triangles*. The orbits and positions of the giant planets are also indicated. From Jewitt *et al.*, 1998 [53].

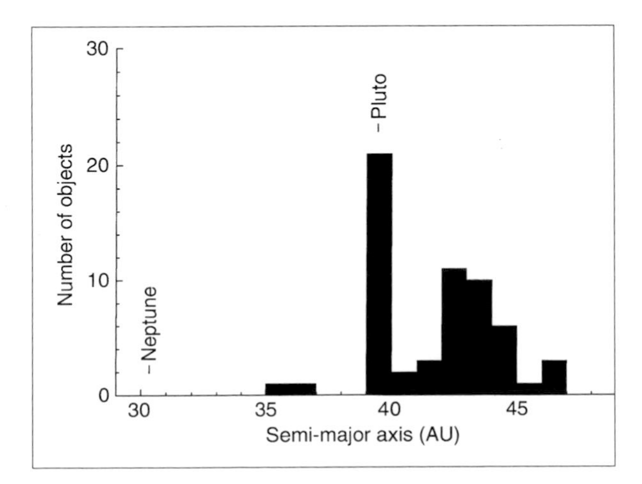

Figure 10.6.
Distribution of semi-major axes for orbits of trans-Neptunian objects.

These objects have visual magnitudes in the range 22 to 25, at the detection limit of today's instruments. Assuming an albedo of 0.04, typical for cometary nuclei and certain distant asteroids, their diameters are estimated to lie between 100 and 400 km. However, they may have higher albedos if they are covered with ices; for example, the albedo of Chiron has been determined as 0.13, and Pluto's as 0.61. In this case, the objects discovered would be smaller. Very little is yet known about these objects. It is difficult to measure their spectra because of their low luminosity. We do know, however, that there is some diversity amongst them. Some are grey, whilst others are significantly redder than the Sun (although less red than Mars). The red colour may indicate a mantle of organic residues on the surface. We do not yet know whether these differences can be attributed to origin, composition, or surface alteration of the bodies.

The distribution of semi-major axes for the orbits of trans-Neptunian objects, shown in Fig. 10.6, reveals peaks. In particular, one third of the objects are grouped with Pluto at about 39 AU. They all have orbital periods around 250 years, i.e. about 3/2 the period of Neptune, with which they are in resonance. Pluto is merely the largest of this class of objects, which some observers have christened the 'Plutinos'. Figure 10.6 shows a peak at around 43 AU, another resonance, corresponding to 5/3 the period of Neptune.

Clearly, the search to date will only have picked out the biggest objects in a population which must be very large. At the present stage of exploration, we can estimate that there must be about two trans-Neptunian objects brighter than magnitude 24 in every square degree of sky close to the ecliptic. Extrapolating, we deduce the existence of about 35 000 objects of diameter above 100 km in the range 30–50 AU from the Sun. This corresponds to about 0.03 of the terrestrial mass. The total mass could reach 0.1 of the terrestrial mass if smaller objects were included. Such a figure would explain the observed flux of short period comets.

It is interesting to ask whether Centaurs and trans-Neptunian objects are related. Their similar sizes and colours suggest that they may belong to the same population. Neptune thus plays the role of guardian in the orbital dynamics of trans-Neptunian objects. Numerical simulations show that certain of these orbits evolve towards orbits of comets in the Jupiter family. According to this hypothesis, the Centaurs would appear to be intermediate objects in such an evolution.

10.2 Relations with other small bodies in the Solar System

10.2.1 Meteors

Showers of meteors (or shooting stars) are a spectacular phenomenon. Although some isolated meteors do exist, most of them occur in swarms. These appear each year at a fixed date, although with variable intensity. Each swarm comes from a fixed direction relative to the stars and is therefore named after the constellation it seems to originate from (see Table 10.1). Meteors are studied visually by counting and recording directions. Radar observations can fix their trajectories and spectral observations of some exceptionally bright meteors have been made.

In 1866, the Italian astronomer Giovanni Schiaparelli (1835–1910) noticed a remarkable coincidence. The direction of one of the more spectacular meteoritic swarms, the Perseids, coincided with the orbit of comet P/Swift–

Table 10.1. Association of the main meteoritic showers with comets and asteroids.

Swarm	Date	Comet or asteroid
Lyrids	21 April	Thatcher 1861 I
η Aquarids	4 May	1P/Halley ?
Bootids	28 June	7P/Pons–Winnecke
N Taurids	30 June	2P/Encke ?
Perseids	11 August	109P/Swift–Tuttle
Draconids	9 October	21P/Giacobini–Zinner
Orionids	20 October	1P/Halley
S Taurids	5 November	2P/Encke
Andromedids	14 November	3D/Biela
Leonids	16 November	55P/Tempel–Tuttle
Ursids	22 December	8P/Tuttle
Sagittarids	Several	(2101) Adonis
Aquarids	January	(1685) Toro
δ Cancrids	15 January	(2212) Hephaistos
S χ Orionids	10 December	(2201) Oljato
Geminids	13 December	(3200) Phaeton

Tuttle which had first appeared in 1862. The orbit of this comet crosses Earth's, reaching its closest approach on 11 August; and the Perseid shower occurs every year around 11 August. The Perseid shower is therefore caused by dust grains released when P/Swift–Tuttle is active and which diffuse along the cometary orbit. They form a cometary trail similar to those found in the infrared by IRAS (see Chapter 6). Comet P/Swift–Tuttle reappeared in 1992. It was thought that its return might cause a spectacular intensification in the Perseid shower. However, the increase was relatively modest, implying that the P/Swift–Tuttle dust trail is rather regularly distributed along its orbit.

Several other meteor swarms have been associated with comets, and these are shown in Table 10.1. As mentioned above, some have also been associated with asteroids, which may have displayed cometary activity in the past. Unlike the P/Swift–Tuttle case, certain swarms are only observed shortly before and after passage of the associated comet. This is the case for the Leonids, related to comet P/Tempel–Tuttle. It has period 33 years and meteors are only visible one year before and after.

Meteors occur when dust particles enter Earth's atmosphere at high speed (around 10 km/s). They heat up to incandescence and vaporise. Dust grains causing this phenomenon measure between a few tenths of a millimetre and several centimetres. Smaller grains are slowed down before they can vaporise and do not produce the same effect. After spending a certain time suspended in the atmosphere, they eventually sediment onto the surface of our planet. These particles are called *micro-meteorites*. The total flux of matter reaching Earth in this way is considerable: several thousand tonnes per year. It is sometimes difficult to distinguish these particles from terrestrial matter (either of natural origin, such as volcanic dusts, or man-produced, such as ash and soot in smoke).

Such particles may be gathered in certain special sites, like the Antarctic ice cap. They have also been collected by stratospheric planes (see Fig. 10.8) and orbiting space stations.

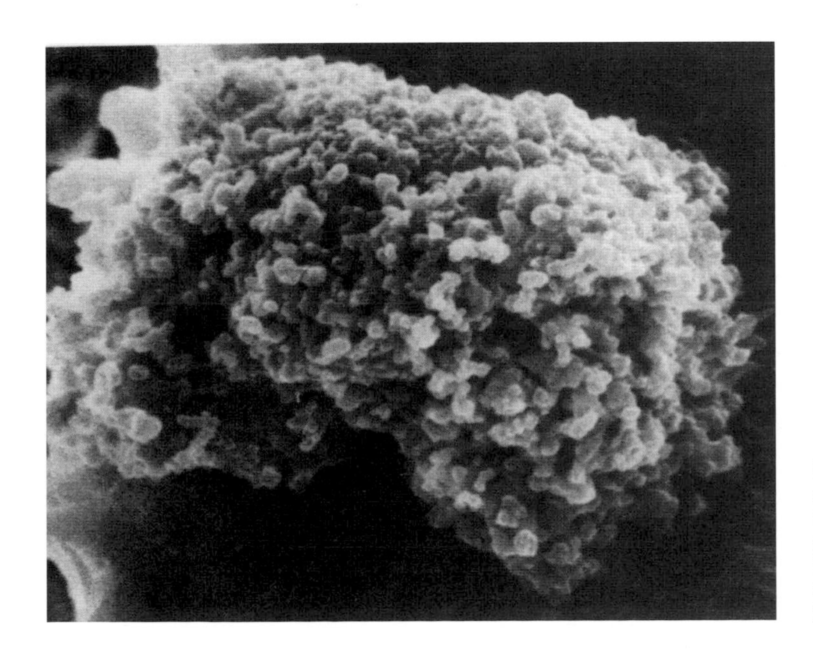

Figure 10.8.
A Brownlee particle collected by NASA using a U2 stratospheric plane. These dust particles are presumed to have cometary origins. They have low density, flaky structure. Some are related to carbonaceous chondrites through their composition. Balsiger *et al.*, 1988 [25].

10.2.2 Meteorites

Bodies of dimensions greater than a few centimetres suffer only surface ablation whilst crossing the atmosphere. These become meteorites. Apart from possibilities opened up recently by space exploration, these are the only macroscopic samples of extraterrestrial matter to reach us. The science of meteorites is highly developed. It is possible to carry out very detailed laboratory analysis of chemical and isotopic composition, as well as physical structure. However, it is not our purpose to go into the details here. In brief, iron, stony-iron and stony meteorites would seem to be examples of highly 'processed' matter. They are closely related to asteroids, in which they may originate. The most numerous are stony meteorites.

The most primitive meteorites are the *carbonaceous chondrites* which are relatively rare. These are composite materials (see Fig. 10.9). They contain *chondrules*, spherical nodules of silicates measuring about 1 mm in diameter, and *bright inclusions* of highly refractory minerals a few millimetres long. These are enclosed in a matrix rich in carbonaceous material (about 3% by mass of carbon) and water (up to 20% crystallisation water). They are friable objects and the least dense of all meteorites (2.0–2.5 g/cm^3). The most famous carbonaceous chondrites, giving rise to the most interesting studies, are from Orgueil (Tarn-et-Garonne, France, 1864), Allende (Mexico, 1969), which weighed more than two tonnes, and Murchison (Australia, 1969). It is common usage to name meteorites after the geographical location of their fall.

In the last century, the chemist Marcelin Berthelot (1827–1907) was able to show that there were organic molecules in the Orgueil meteorite. Recent analyses of carbonaceous chondrites have revealed an extraordinarily rich panoply of organic molecules, including hydrocarbons, alcohols, acids, benzene and polyaromatic compounds, and even, it would seem, amino acids.

Could there be a relation between carbonaceous chondrites and cometary nuclei? Clearly they are not cometary nuclei, since they contain no ices or volatile elements. Chondrules could only have formed after solidification at several hundred kelvins, and this is much higher than temperatures which prevailed during formation of cometary nuclei.

10.3 Formation of the Solar System

The Sun, planets and smaller bodies of the Solar System formed 4.7 billion years ago by condensation of an interstellar cloud. This model, based on the idea of a primitive nebula, was originally proposed by Kant (1724–1804) and Laplace (1749–1827) and is now almost unanimously accepted by astronomers. However, the details remain to be filled in. In particular, the origin of comets has not yet been completely elucidated (see Figs. 10.10 and 10.11).

In the first stage, an interstellar cloud collapsed under its own weight. The process was probably triggered by a gravitational instability. This produced a massive central component, the protosun. Collapse would have been accompanied by heating, as gravitational energy was converted into heat. When the density and temperature of this central region reached a critical threshold, thermonuclear reactions could begin. In this stage, hydrogen and helium were

Figure 10.9. opposite
Carbonaceous chondrite. A thin slice of the Allende meteorite photographed in polarised light. The circular pattern is a chondrule (a spherical sililcate grain) of diameter 2.5 mm. Pieces of silicate, sulphur and nickel-iron are also visible. Black regions are made up of a matrix of very fine silicate and carbonaceous grains. Lewis and Anders, 1983 [61].

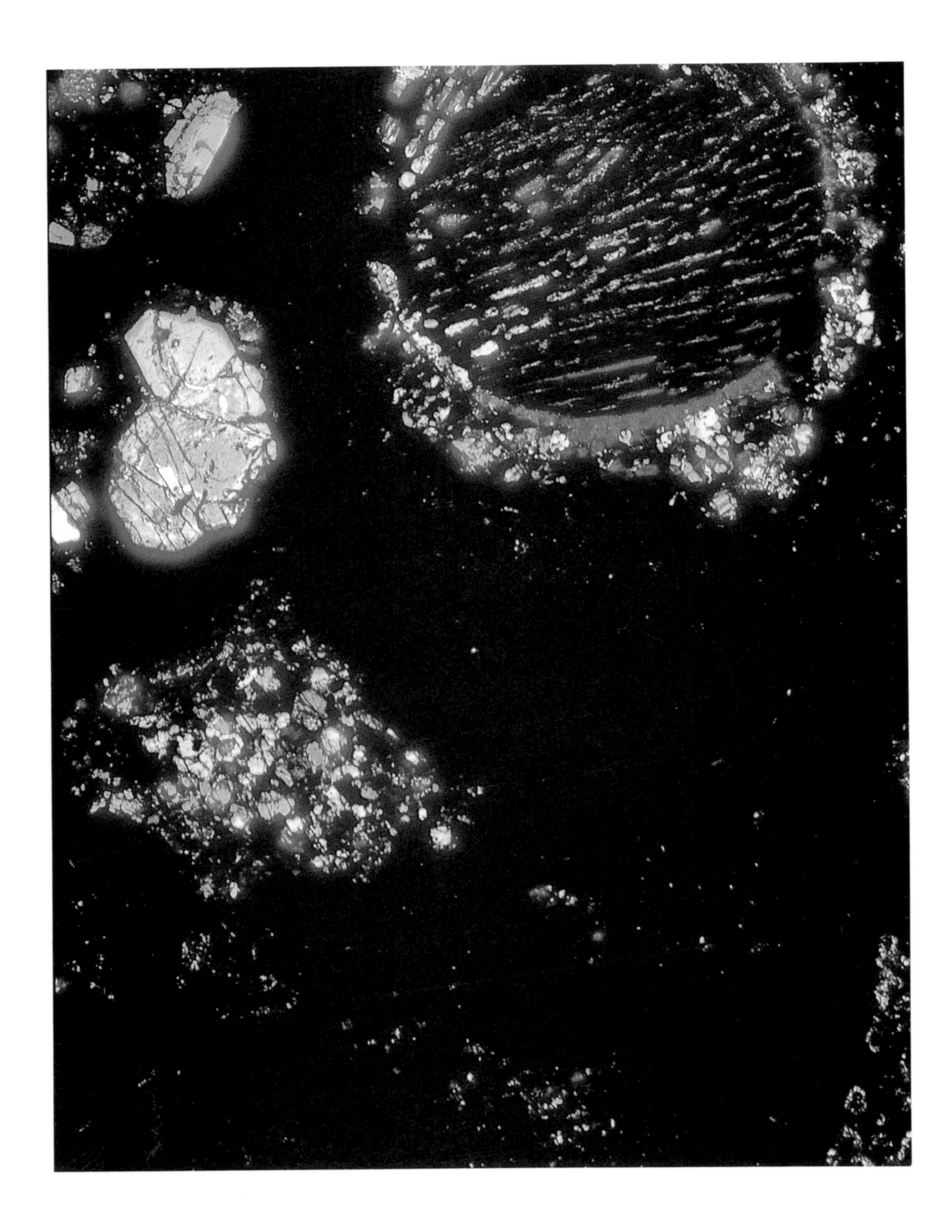

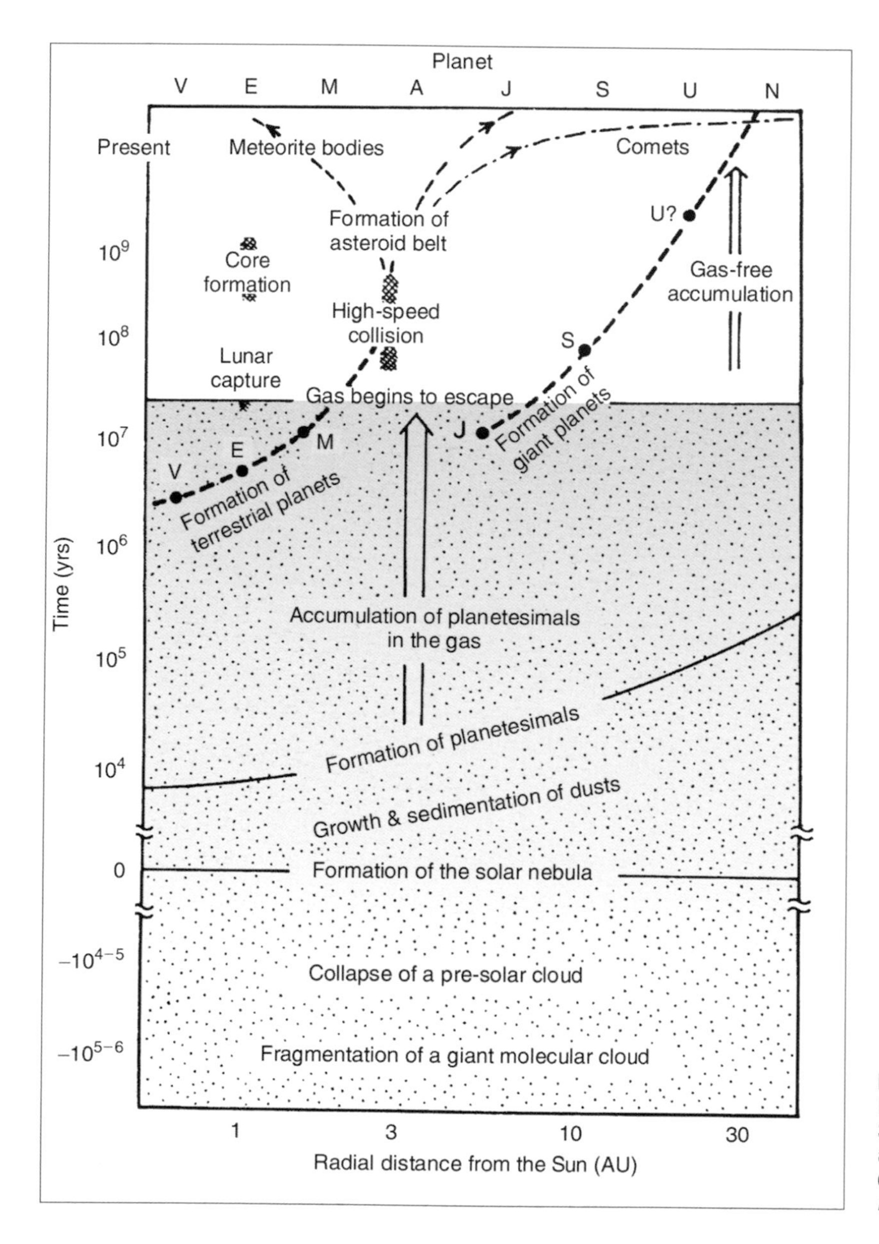

Figure 10.10.
Diagram showing the formation of the Solar System. The various processes are positioned relative to heliocentric distance (*abscissa*) and time (*ordinate*). From Hayashi *et al.*, 1985 [51].

converted into heavy elements and vast quantities of energy released. The central condensation became a luminous star, the Sun. At the same time, outer regions of the interstellar cloud also began to collapse. However, because of angular momentum conservation, this stage of collapse led to formation of a rotating disk, the protoplanetary disk.

At this point, the various scenarios diverge. According to the 'classic' model upheld by the American Alastair Cameron, the protoplanetary disk would have fragmented into rings. These then condensed to form the various planets in the same way as the central condensation formed the Sun. More recent theories, such as those proposed by the Russian V. Safronov and

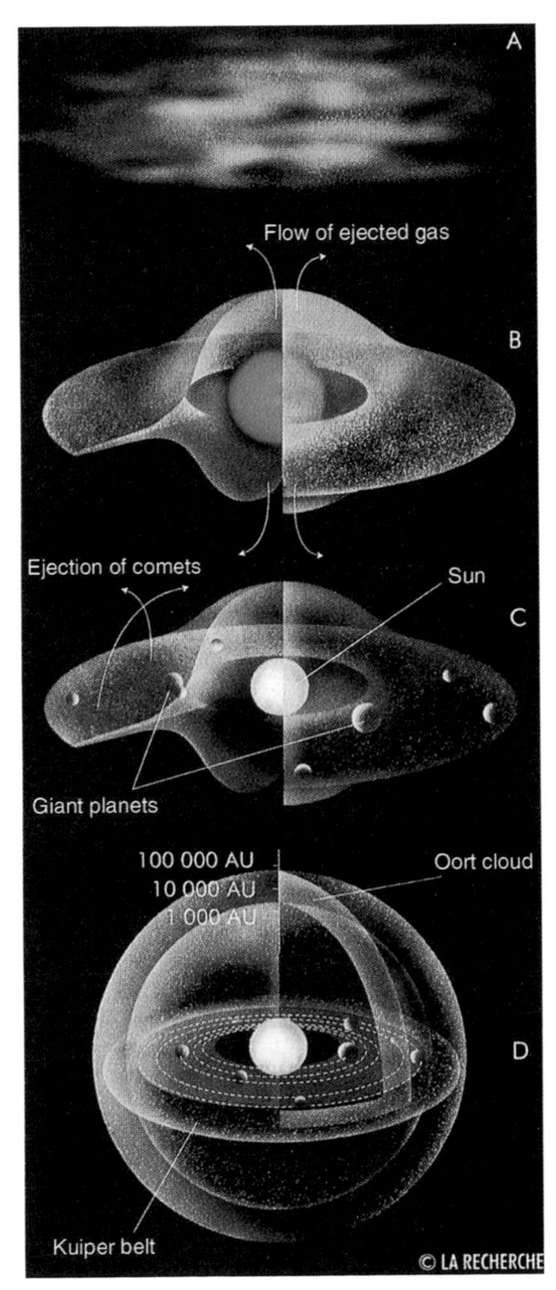

Figure 10.11.
Formation of the Solar System, the Oort cloud and the Edgeworth–Kuiper belt. Bockelée-Morvan and Crovisier, 1994 [30]. Copyright 1994, La Recherche, Paris, France.

the Japanese at Kyoto University, bring in an additional stage, the planetesimal stage. Planetesimals are bodies resulting from aggregation of dust and gas, producing objects with dimensions anywhere between a few hundred metres and a few kilometres. This aggregation occurs through collisions, and relative speed of colliding bodies is a critical parameter. In very high speed collisions, bodies are destroyed, but if the speed is too low, the frequency of collisions is also low and growth of planetesimals correspondingly slower. Calculations show that this must have been a relatively short stage, lasting only about 100 000 years.

It was during a later stage which lasted much longer, about a hundred million years, that planetesimals would have formed planets and the larger asteroids known to us today. This was also the result of agglomeration by collisions. In this stage, an essential role would have been played by gravitational interaction between these massive bodies. In the case of the giant planets, it allowed them to accrete gas left over from the protoplanetary nebula. It also effected a ruthless selection among any remaining planetesimals. They would have struck some larger body, been torn apart by its tidal forces, or been ejected towards the outer Solar System. In the latter case, they could either contribute to the Oort cloud or escape completely from the Solar System. Only those planetesimals located in stable orbits could remain in the inner Solar System, e.g. asteroids in the Main Belt. Residual gas from the protoplanetary nebula would have disappeared quickly, swept away by the Sun's stellar wind. According to Safronov, gases would have been removed at the very beginning of planetesimal formation; or according to the Kyoto school, slightly afterwards. Gases would play a key role in slowing down small planetesimals, allowing them to take part in low speed, non-destructive collisions.

In this scenario, comets are planetesimals. We have seen that planetesimals formed in the inner Solar System would have been cast out to great distances. Only the trans-Neptunian objects would have remained in the region where they formed.

The temperature of the protoplanetary disk decreased from the centre outwards. As a result, planetesimals formed at different distances would not have the same composition (see Fig. 10.12). Thus, cometary ices observed today could only have condensed several tens of AU from the Sun. But great differences in composition are to be expected depending on the distance at which the comet condensed. Study of cometary composition is only just beginning to reveal the chemical diversity of these objects. But this may be the key to understanding their origin, and it might teach us about the formation of the Solar System itself.

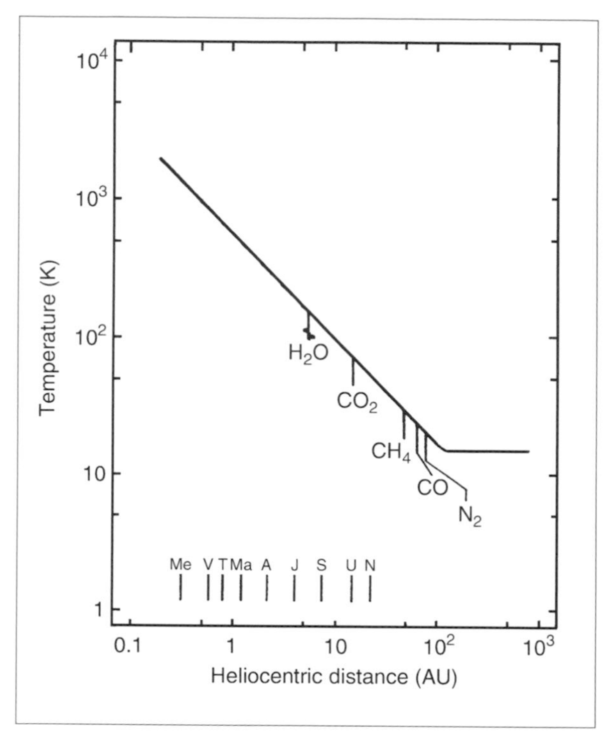

Figure 10.12.
Temperature of the protosolar nebula and equilibrium temperatures for sublimation of various ices. From Yamamoto, 1985 [82].

β Pictoris and other planetary systems

Another way of studying the formation of our Solar System is to search for other planetary systems at present developing elsewhere in the Universe. Present technology is still insufficient to detect potential planets by direct imaging, even those gravitating around the closest stars. Such systems are therefore sought by indirect means. One approach is to observe tiny periodic displacements, or velocity oscillations, of certain stars, which might betray the presence of one or more planetary companions. Another is to study light curves of stars, where brightness variations might be due to eclipse by an invisible companion. Several planets have been discovered in this way around nearby stars. All of them are giant planets of size and mass comparable to Jupiter. Terrestrial planets remain undetectable with such techniques. However, an analysis of frequency anomalies in the millisecond binary pulsar PSR 1257 + 12 has revealed that Earth-mass planets could be orbiting around this pulsar; but pulsars, which are formed through an extreme step in stellar evolution, cannot be the centre of solar systems comparable to ours. On the other hand, dust disks have been discovered around certain nearby stars.

T Tauri stars are very young stars, still interacting with their accretion disk. They are characterised by a large *bipolar flow* of matter escaping from their poles, by continued accretion of matter from the circumstellar disk and by intense chromospheric activity. The latter is due to the fall of matter being captured by the star. Such objects probably correspond to the first stages of stellar and planetary formation.

Further evolved systems are the stars β Pictoris (see Fig. 10.13), α Lyrae (Vega) and α Piscis Austrinus (Fomalhaut). IRAS detected a strong infrared excess emission coming from these objects. They are all close, at around 10 parsecs (1 parsec = 3.086×10^{16}m = 3.26 light-years), and their spectral types indicate an age of about a billion years. Infrared excesses have been attributed to dust disks extending out as far as 100 or 1000 AU from the stars. These disks also comprise an empty central zone, several tens of AU in diameter. Such stars have gone beyond the age of planetary formation. Their disks may be the remains of protoplanetary disks, equivalent to our Edgeworth–Kuiper belt or the Oort cloud. In the case of β Pictoris, an optical image of the disk has been obtained by stellar coronagraphy (Fig. 10.13).

Spectroscopic observation of these disks has shown that there is still some gas present. In addition, absorption lines of metallic atoms appear randomly, with frequency shifts indicating velocities of several hundred kilometres per second relative to the central star. This rather surprising observation has been explained as the result of small bodies falling towards the central star and volatilising, like sungrazing comets. ISO observations of such disks have shown that there is a great similarity between their dust and cometary dust, with the presence of magnesium-rich olivine (forsterite, see Fig. 6.6).

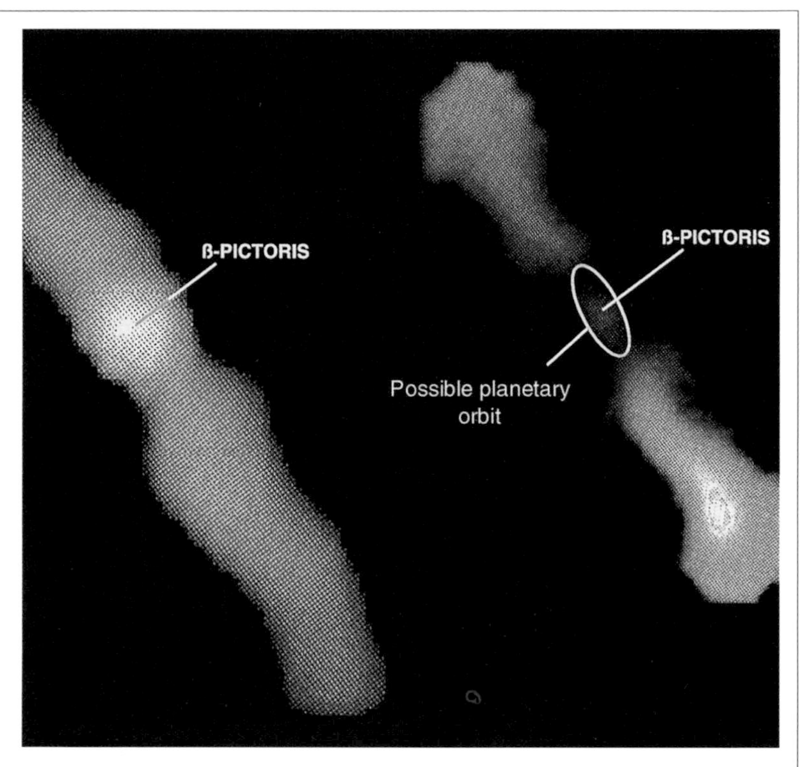

Figure 10.13.
Infrared image (at 10 μm) of the disk around β Pictoris, taken with the 3.6 m ESO telescope. The raw image is shown on the *left*. On the *right*, the star has been subtracted, together with temperature effects which are more marked in the centre than at the edges. The image brightness then corresponds to the quantity of dust. An empty central region is clearly apparent. It is attributed to dust ejections following gravitational perturbations by one or more planets. From Pantin and Lagage, 1995 [73].

11 The future of cometary research

During the past fifteen years, cometary research has made a tremendous leap forward. Following the Halley campaign which mobilised every available means of ground-based astronomy and space research between 1985 and 1986, several bright comets have made their appearance. There have been apparitions of both periodic and non-periodic comets: Wilson 1987 VII, P/Brorsen–Metcalf, Austin 1990 V, Levy 1990 XX and P/Swift–Tuttle are a few examples. Another comet, P/Grigg–Skjellerup, was the flyby target of a space probe in 1992. The collision of comet Shoemaker–Levy 9 with Jupiter in July 1994 was observed by an exceptional campaign which coordinated telescope programmes and space observations on an international scale. And finally, two exceptionally bright new comets appeared at the end of the century: Hyakutake in 1996 and Hale–Bopp in 1997. Complete analysis of all the data will no doubt take several years.

Today, then, astronomers possess new information to develop their understanding of comets and solve some of the open problems. At the same time, they have acquired new tools, such as infrared and millimetre spectroscopy, faint object imaging, and *in situ* mass spectroscopy.

The experience of recent years has shown that ground-based astronomy, with its new high performance instruments, can and must remain a rich source of information. This is particularly true for exploratory research (e.g. the search for objects in the Edgeworth–Kuiper belt) and statistical research (e.g. concerning chemical composition and relative abundances in the various classes of comets). Space astronomy in Earth orbit will provide invaluable support for such research at a distance, opening the way to wavelengths inaccessible from the ground. And finally, it is becoming clear that future stages in cometary research will require *in situ* analysis of a cometary nucleus. The ultimate stage in this exploration will be return of cometary samples. These can be studied at leisure in the laboratory, in the same ways as lunar and meteoritic samples.

11.1 Ground-based telescopic observation

In the visible and infrared regions, recent progress has affected the size of telescopes and also the sensitivity and dimension of cameras. It is now possible to follow the activity of distant comets. In addition, there is a systematic search programme for objects in the Edgeworth–Kuiper belt which is beginning to obtain results. Likewise, progress has been achieved with cooled infrared spectrometers. This means that activity and chemical composition of faint cometary nuclei can now be studied. This type of programme is beginning to benefit from very large telescopes in the 10 m category. The Keck (named after an American sponsor who helped to finance it) consists of two brand new 10 m telescopes at the summit of Mauna Kea in Hawaii. The VLT (Very Large Telescope) will group together four 8 m telescopes in Chile at the ESO (European Southern Observatory). One of these is already in service.

The Spaceguard project envisaged by NASA may become an international project. Its aim is to record all asteroids of a certain size whose orbit intersects Earth's orbit and which might pose a threat to our planet. Several wide field telescopes of diameter 2.5 m and equipped with large format CCD cameras will be entirely devoted to this purpose. An important and difficult part of the project is the development of software for automatic detection of moving objects. A spin-off from such a project will be the detection of many comets. Similar programmes specifically aimed at cometary searches may well be developed. Indeed, asteroid searches will be restricted to surveillance of ecliptic regions, so that comets with highly inclined orbits would escape detection.

When such projects have become a reality, we may wonder whether amateur astronomers will continue to play any role, in competition with these well-trained comet-hunting robots. The amateurs do have numbers on their side, allowing rapid surveillance of the whole sky. It seems quite likely that they will continue to discover most bright comets outside the ecliptic, unless a great number of automatic observing stations can be put into operation.

Another technique which should prove fruitful for cometary observation in the visible and near infrared is the recent method of adaptive optics. A deformable secondary mirror is connected into a feedback loop in such a way as to restitute the diffraction pattern of the telescope and counter atmospheric turbulence in real time. Spatial resolution in cometary imaging can thereby be improved by a factor between 3 and 10. This will considerably advance the study of cometary circumnuclear regions.

The submillimetre radio region is also a promising part of the spectrum. The relevant wavelengths lie between a few hundred micrometres and one millimetre. We have already seen the successes of heterodyne spectroscopy in searching for parent molecules. Present technological developments suggest that within a few years we may have receivers capable of detecting a frequency of 1 THz. This corresponds to a wavelength of 300 μm. The advantages of such an innovation are two-fold. Firstly, the field of view will be reduced, allowing better spatial resolution. Secondly, the range of observable frequencies will be extended. It will be possible to search for new parent molecules or use more intense transitions with a correspondingly increased sensitivity.

An international project, the ALMA project, being developed at the moment aims to construct a giant interferometer operating in the submillimetre region. The instrument would comprise about fifty medium-sized antennas with total area around 7000 m^2. An instrument of this kind would combine the angular resolution of an interferometer with the sensitivity of a single large antenna. It could be set up in a high altitude site in the southern hemisphere (e.g. Chile). A large number of short period comets would become accessible.

11.2 Observation from Earth orbit

Observations of comets from vehicles orbiting the Earth have already proved their worth. During almost 20 years, the IUE satellite (International Ultraviolet Explorer) has produced

Storage and dissemination of data

Maybe even more than any other branch of astronomy, the study of comets requires accumulation of observational data over long periods of time. This enables cometary scientists to establish the long term evolution of such objects and collect large enough samples for statistical study. Information about past comets gathered from the ancient chronicles (either in the West or from China) is of great importance in this respect. We must in turn ensure that the whole range of today's observations, from the simple magnitude estimations of amateurs to images obtained by space probes, can be transmitted in their entirety to future generations.

Systematic, exhaustive and reliable data storage is needed. It must allow our successors to understand the conditions in which observations were made, to see the results of those observations and, if necessary, to reprocess the data using new methods. The Halley campaign provides a good example of the kind of effort required. The results have been archived on twenty-six CDROMs. A similar effort was made for observations of the collision between Jupiter and SL9. Orbit catalogues and compilations of astrometric observations relating to other comets are kept up-to-date by the Central Bureau for Astronomical Telegrams of the IAU. A record of visual magnitude observations, mainly provided by amateur astronomers, is maintained by the International Comet Quarterly group. Unfortunately, there does not yet exist a similar organisation for storing cometary spectra and images on an international scale, although some 'anthologies' have been published on various occasions.

A great number of cometary observations have to be made with little warning. A further specific need is thus the rapid dissemination of information: discovery of new comets, calculation and updating of orbital elements, reports of evolution and sometimes remarkable unexpected events that may occur (e.g. bursts, appearance of jets, nuclear break-up). An efficient solution is provided by improvements in communications, and particularly the development of computer networks and electronic mail. Rapid dissemination of information was for a long time limited to telegram and telex. During the Kohoutek observation campaign in 1973–74, NASA set up a telex information service. The Halley campaign coincided with the first computer networks. A hotline was set up by the International Halley Watch and many astronomers (although not those in the Third World or Eastern Europe) were able to consult the latest records or announce their own results instantaneously.

A new step was taken during observations of the collision between SL9 and Jupiter. This was the *mail exploder* which allowed information to be sent out instantaneously to the whole observing community the moment it was obtained. Several hundred messages were thus sent out during the week of impacts, sometimes saturating subscribers' computers.

Finally, all this information is now freely available, almost in real time, on the World Wide Web. Several key internet addresses are given in the Bibliography section.

a very complete database of ultraviolet cometary spectra.

More recently, the Hubble Space Telescope, repaired at the end of 1993, has continued this work in a quite spectacular way. Fragments of comet Shoemaker–Levy 9 were imaged with a quality never before achieved. It then followed the various stages in their collision with Jupiter in July 1994, imaging the traces left on the planet over several months afterwards. Ultraviolet spectra have also been recorded. These reveal certain cometary compounds and auroral phenomena. Hubble imaging and spectroscopy has also contributed a great deal to the study of comets Hyakutake and Hale–Bopp.

A successor for the Space Telescope is at present under study. This is the NGST (New Generation Space Telescope). It would have an 8 m mirror and would be cooled, thereby allowing observations in the mid-infrared.

Space observatories, although not originally designed for cometary observation, have proved to be essential tools in cometary science. We have already seen the example of the solar coronagraph (see Section 7.3.3) and the X-ray telescope (see Section 6.3). Cometary observation has since become a standard part of the programme of these instruments.

During the two and a half years of its activity (1996–98), the ISO (Infrared Space Observatory) provided a unique opportunity for

Figure 11.1.
The Hubble Space Telescope after it was separated from the shuttle, on 25 April 1990. Photo NASA/ESA.

cometary research (see Section 6.1.5). Luck was with cometary scientists when the exceptional comet Hale–Bopp appeared during the period of operation of this satellite. Several short period comets were also observed by ISO in spectroscopic mode, in particular 22P/Kopff and 103P/Hartley 2; and distant comets were observed by imaging and photometry. The ISO cometary observation programme may be continued, at least in part:

* using the infrared equipment of the Space Telescope;
* by SIRTF (Space Infrared Telescope Facility), a new infrared satellite which NASA hopes to launch in 2001;
* by SOFIA (Stratospheric Observatory for Infrared Astronomy), a joint project between the USA and Germany. This consists of a 2.5 m infrared telescope to be flown by plane at altitudes over 12.5 km. It may also go into operation in 2001.

Another interesting area of research in cometary physics is space exploration in the submillimetre region. Several projects are under study. Following operation of ISO, this is the last region remaining unexplored in the electromagnetic spectrum. A Swedish project named ODIN, after the god in Scandinavian mythology, will be run jointly with France, Canada and Finland. It aims to put an antenna of collecting area approximately 1 m^2 into orbit in 2000. It will be equipped with heterodyne receivers in different channels corresponding to water and oxygen transitions. A similar satellite, the SWAS (Submillimeter Wave Astronomy Satellite) was put into operation by NASA at the end of 1998.

Even further into the future, the FIRST project (Far Infrared and Submillimetre Space Telescope) is one of the major projects at the European Space Agency. It has a 3 m antenna and comprises a series of coherent detection instruments (heterodyne receivers), together with incoherent detection instruments (Fabry–Pérot spectrometers). This ambitious project aims to cover the whole submillimetre region from 500 GHz to more than 1 THz with the heterodyne technique. This corresponds to resolving power of 10^6. It also aims to cover the far infrared from 85 to 300 μm with resolving power 10^3 to 10^4. The heterodyne instrument in particular will be of great interest for cometary research. Such high resolving power is especially well suited to analysis of cometary lines which are always very narrow. The vast spectral coverage will allow a systematic search for parent molecules. Since detectors will be cooled by active cryogenic units, they should have a lifespan of at least five years in terrestrial orbit. However, we must be patient: the FIRST mission is planned for launch in 2007.

11.3 *In situ* cometary exploration

For several decades now, cometary scientists have been convinced of the need for close-up investigation of a cometary nucleus, if the mysteries of cometary composition are to be penetrated. The success of the Halley space exploration has only strengthened this conviction. How else could we unambiguously identify the hydrocarbons in the nucleus? How could we understand why the nucleus is so dark? How could we obtain a precise determination of the mass and density of cometary nuclei? Or the porosity and structure of their matter? How could we discover which ices are contained within the nucleus? There seems to be only one solution: a space mission which, instead of just flying by as in the Halley case, would follow the comet from aphelion to perihelion, analysing the whole development of its activity.

Such a mission could not at the present time be organised for an unexpected new comet, although its nucleus would in that case be of great interest. We must limit ourselves to a short period comet of well-known orbit. Likewise, a comet whose orbit was highly inclined with respect to the ecliptic would also have to be ruled out; it would require too much energy to project a space probe out of the ecliptic plane. Energy can be saved by using the technique of *gravity-assisted manoeuvres*. The probe is sent close to Venus, Mars or Earth in order to modify its trajectory without extra cost. All these considerations seriously limit the number of possible targets for such a mission. About ten comets in the Jupiter family are potential candidates.

Another technique should help future cometary probes to attain their targets more easily, and also more rapidly. This is *ion propulsion*, also known as *solar electric propulsion*. The process uses a rocket whose 'fuel' is made up of ions accelerated to high speeds in an electric field. Energy for both ionisation and electric field is provided by solar panels. The thrust is very weak and intended to operate over a long period (weeks or months). The advantage of this technique over traditional chemically-fuelled rockets is its very great efficiency for a given mass of 'fuel'. The first trials, or technology flights, should be made in NASA's DS1 (Deep Space 1) missions and the ESA's SMART mission to comets or asteroids.

Table A.3, entitled *Space missions*, given in the Appendix sums up missions to comets and asteroids already carried out, together with those at present under study. Several cometary missions are under construction at NASA and the ESA. On the American side, following abandonment of the CRAF project (Comet Rendezvous and Asteroid Flyby) in 1992, NASA first concentrated its attention on a flyby of nearby asteroids called NEAR (Near Earth Asteroid Rendezvous). Launched in February 1996, the NEAR probe flew by the asteroid Mathilde on 27 June 1997 and moved towards the asteroid Eros which it should encounter in February 2000.

More relevant to cometary research is the American mission Stardust. This mission aims to bring a sample of cometary dust back to Earth. The payload includes the dust collector, a mass spectrometer, an impact analyser, a navigation camera and a radio wave instrument. The objective is to collect a thousand cometary particles of size greater than 15 μm, as well as a hundred interstellar particles of size greater than 0.1 μm. It was launched in 1999 and an encounter with comet 81P/Wild 2 is planned in 2004. The probe should fly by at a distance of 100 to 200 km. It will then collect cometary samples and return to Earth in 2006, collecting interstellar particles on the way back.

The major cometary research project at the beginning of the next century will be the Rosetta mission at present under development at the ESA. The name refers to the Rosetta stone which helped Champollion to decipher Egyptian hieroglyphics. It is hoped that data from this mission concerning cometary nuclei will have a similar effect in helping us to understand the history of the Solar System. An in-depth exploration of a comet is planned, during the whole period of its approach towards perihelion, the aim being to understand the nucleus, identify its volatile and refractory elements, and analyse how cometary activity evolves. The Rosetta mission comprises an orbiter and a probe intended to make a soft landing on the cometary nucleus itself. The operation will be rather delicate, given the comet's almost zero gravitational field.

The Rosetta orbiter will be equipped with traditional remote sensing instruments: cameras, spectrometers operating in the ultraviolet, visible and infrared, and even a small radio telescope. It will also carry instruments for *in situ* measurements in the coma: mass spectrometers and dust analysers. Several instruments devoted to plasma physics are also planned. The lander probe will have a lifespan of between a few days and several weeks. Its instruments will carry out *in situ* analysis of cometary material: mass spectrometry, X-ray and gamma-ray spectrometry, etc.

Which comet will be the target of the Rosetta mission? It must not be too active, since dust jets could damage instruments. But it must be sufficiently active to produce phenomena worth observing! Comet 22P/Wirtanen has proved to be a suitable compromise. Two asteroids will be flown by during the journey to the comet. The details of the mission remain to be established. The experts still have some time available, since launch is planned for 2003.

According to the scenario favoured at the present time, Rosetta will leave in January 2003 and reach comet P/Wirtanen 8 years later, in August 2011 (see Fig. 11.3). Apart from the more delicate manoeuvres when it will pass close to Mars and Earth, and during asteroid flybys when it will undertake studies, the probe will remain in a state of hibernation for the main part of this journey. When it meets the comet, the probe will be at about 5 AU from the Sun.

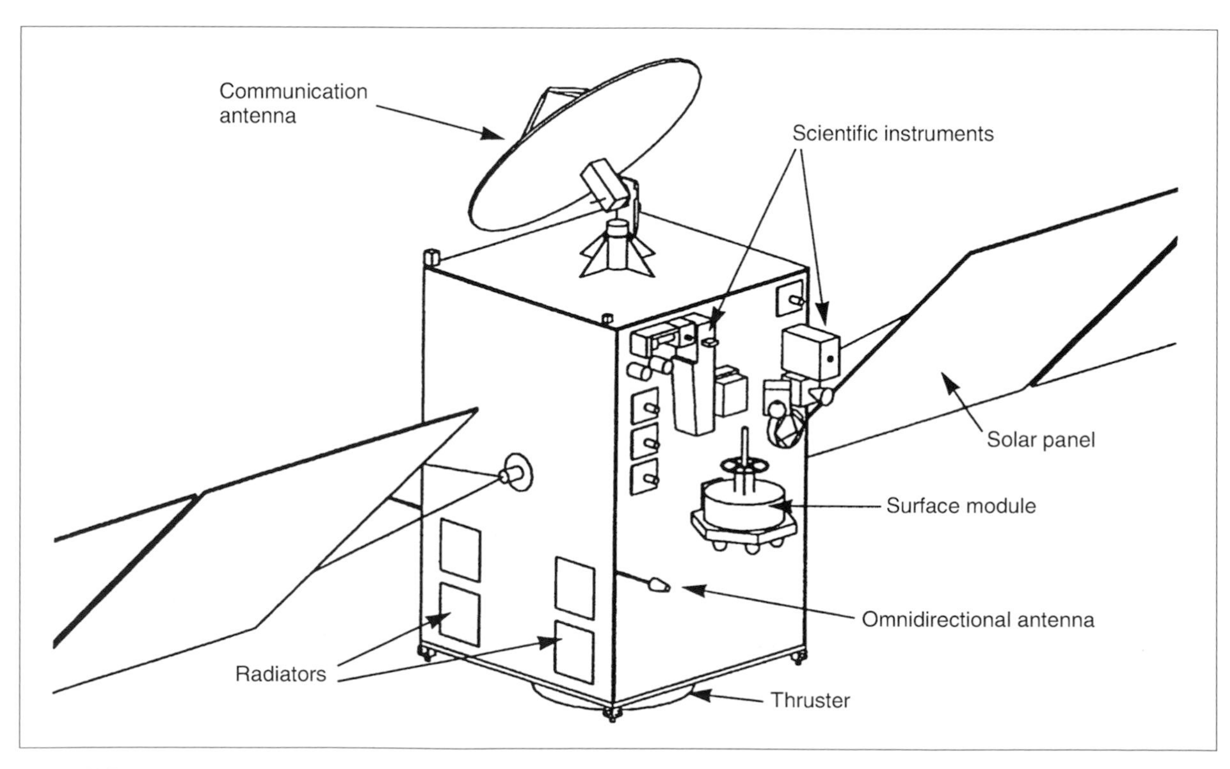

Figure 11.2.
The Rosetta probe. From an ESA document.

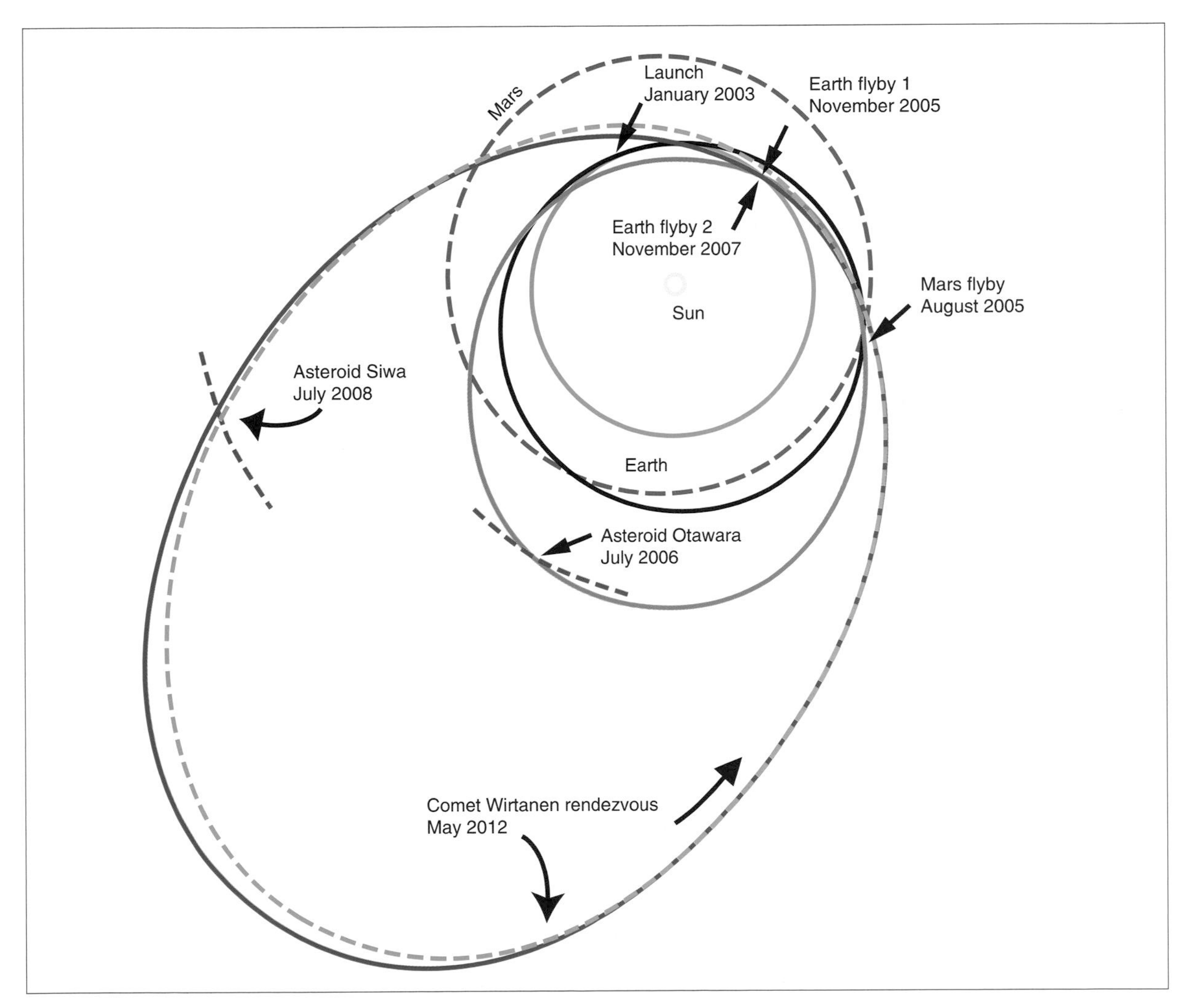

Figure 11.3.
Trajectory of Rosetta on its way to the encounter with comet P/Wirtanen. From an ESA document.

It will be unable to make observations at this stage since it obtains its energy from solar panels and they will not provide enough energy at such a distance from the Sun. Approach manoeuvres will only begin several months later. Then, when comet and probe are less than 4 AU from the Sun, the probe will begin to survey the cometary nucleus. To do so, it will go into orbit around the nucleus (eccentric orbit, semi-major axis between 10 and 150 km). The aim of this first exploratory phase is to find suitable landing sites for the lander. The latter will be sent down when the comet is about 3 AU from the Sun (around October 2012). At this time, the comet will not be very active. Subsequently, the orbiter will study how its activity evolves up to perihelion (July 2013, at 1.06 AU from the Sun), and possibly beyond.

Rosetta's first task is therefore to study the nucleus from a distance as a preparation for *in situ* examination by the lander. It will have to determine its dimensions, shape, mass, density and rotational state. This will be done by cameras equipped with filter sets, which can operate as wide band photometers. An infrared imaging spectrometer will draw up spectral maps of the nuclear surface and specify its mineralogical composition. Radio and infrared radiometry of the nucleus will ascertain its temperature. Radio wave radiometry will even measure the temperature a few centimetres below the surface, making it possible to follow how it heats up and cools down. The aim, of course, is to distinguish icy regions which may become active from regions covered over by a dust mantle or refractory organic residues.

Mass spectrometers will perform better than those aboard Giotto and Vega. These had resolutions only just above 100 and were unable to separate isobars, nuclei with the same mass but different atomic numbers. A resolution of about 1000 would remove certain ambiguities. It could distinguish OH from NH_3 and CH_3OH from S, for example. A resolution above 2500 is needed to separate N_2 from CO and ^{13}C from ^{12}CH.

An infrared spectrometer of resolution 2000, in addition to the low resolution imaging spectrometer, should identify a great many compounds in the nuclear surface as well as in the cometary atmosphere. It should resolve the rotational structure of vibrational bands, thereby revealing physical conditions prevailing in the first few kilometres of the cometary atmosphere.

A small radio telescope of diameter 30 cm, equipped with spectrometer and operating in the submillimetre region, will study a selection of water, carbon monoxide, methanol and ammonia lines. It may detect the first signs of cometary activity and measure the composition, velocity and temperature of gas jets.

The lander will be equipped with probes for measuring temperatures at various times and depths, so that the thermal properties of the nucleus can be studied. It will also measure permittivity as an aid to interpreting remote sensing by radio wave radiometry, and it will measure mechanical strength. Small cameras will observe surface structure and morphology. X-ray and gamma-ray spectrometers will determine which elements make up nuclear material. Chemical analysis will be carried out by a pyrolytic device in order to ascertain the molecular composition of refractory compounds in the outer layer.

The internal structure of the nucleus will be studied by propagating radio waves through it, between an emitter on the lander and a receiver on the orbiter. This uses a technique analogous to tomographic scanning in medicine.

All these missions are without doubt necessary steps in the study of comets and the history of the Solar System. However, billions of small bodies remain to be explored with more traditional means, and we feel sure that their great diversity will continue to surprise us.

Appendix

Table A.1. Famous comets.

Comet (a)		(b)	(c)	Perihelion date	q (AU)	e	P (years)	m_{max}	H_0	Remark
C/1680 V1	'Kirch's comet'	1680		18 Dec. 1680	0.0062	1.000		2	4	Sungrazer (Kreutz family)
1P/1682 Q1	Halley	1682		15 Sep. 1682	0.58	0.968	77.4	2	4	
C/1729 P1	'Sarabat's comet'	1729		16 Jun. 1829	4.05	1.0		4	−3	Intrinsically very bright
C/1743 X1	'Chéseaux's comet'	1744		1 Mar. 1744	0.222	1.0		−4	0.5	
1P/1758 Y1	Halley	1759 I		13 Mar. 1759	0.58	0.968	76.9	0	4	Return predicted by E. Halley
D/1770 L1	Lexell	1770 I		14 Aug. 1770	0.67	0.786	5.6	2	8	Approached Earth at 0.015 AU
3D/1772 E1	Biela	1772		17 Feb. 1772	0.99	0.726	6.9	6	7	Nucleus split in 1846 – now lost
2P/1786 B1	Encke	1786 I		31 Jan. 1786	0.34	0.848	3.3	5	9	Comet of shortest period
C/1811 F1	Great comet	1811 I		12 Sep. 1811	1.04	0.995		0	0	Famous for its wine!
1P/1835 P1	Halley	1835 III		16 Nov. 1835	0.59	0.967	76.3	2	4.5	
C/1843 D1	Great March comet	1843 I		27 Feb. 1843	0.0055	0.999	51.3	−7	5	Sungrazer (Kreutz family)
4P/1843 W1	Faye	1843 III		17 Oct. 1843	1.69	0.556	7.4	6	5	
C/1858 L1	Donati	1858 VI		30 Sep. 1858	0.58	0.996		1	3	
C/1861 J1	Great comet	1861 II		12 Jun. 1861	0.82	0.985	409	0	4	
109P/1862 O1	Swift–Tuttle	1862 III		28 Aug. 1862	0.96	0.963	131	2	4	Associated with Perseid meteors
C/1864 N1	Tempel	1864 II		16 Aug. 1864	0.91	0.996		2	6	First spectral observations
55P/1865 Y1	Tempel–Tuttle	1866 I		11 Jan. 1866	0.98	0.906	33.5	5	9	Associated with Leonid meteors
C/1868 L1	Winnecke	1868 II		26 Jun. 1868	0.58	1.0		5	8	
C/1874 H1	Coggia	1874 III	1874c	9 Jul. 1874	0.68	0.998		1.5	6	
C/1881 K1	Great comet	1881 III	1881b	16 Jun. 1881	0.73	0.996		1.5	4	
C/1882 R1	Great September comet	1882 II	1882b	17 Sep. 1882	0.007 75	0.9999	759	−5	1	Sungrazer (Kreutz family)
C/1887 B1	Great southern comet	1887 I	1887a	11 Jan. 1887	0.0048	1.0		2	6	Sungrazer (Kreutz family)
16P/1889 N1	Brooks 2	1889 V	1889d	30 Sep. 1889	1.95	0.471	7.1	9	7	Disrupted by Jupiter
C/1901 G1	Great comet	1901 I	1901a	24 Apr. 1901	0.245	1.0		0	6	
C/1907 L2	Daniel	1907 IV	1907d	4 Oct. 1907	0.512	0.999		2	4	
C/1908 R1	Morehouse	1908 III	1908c	26 Dec. 1908	0.945	1.0007		5	4	
C/1910 A1	Great January comet	1910 I	1910a	17 Jan. 1910	0.129	0.9999		−5	5	
1P/1909 R1	Halley	1910 II	1909c	20 Apr. 1910	0.59	0.967	76.1	2.5	4.5	
C/1911 O1	Brooks	1911 V	1911c	28 Oct. 1911	0.49	0.997		2	5	
7P/	Pons–Winnecke	1927 VII	1927c	21 Jun. 1927	1.04	0.686	6.0	4	11	Approached Earth at 0.039 AU

C/1927 X1	Skjellerup–Maristany	1927 IX	1927k	18 Dec. 1927	0.176	0.9998		–6	5	
73P/1930 J1	Schwassmann–Wachmann 3	1930 VI	1930d	14 Jan. 1930	1.01	0.672	5.4	7.5	12	Approached Earth at 0.062 AU
C/1940 R2	Cunningham	1941 I	1940c	16 Jan. 1941	0.368	1.0005		3.5	6	
C/1947 X1	Southern comet	1947 XII	1947n	2 Dec. 1947	0.110	0.9995		0	6	
C/1948 V1	Eclipse comet	1948 XI	1948l	27 Oct. 1948	0.135	0.999 94		2	5.5	
C/1956 R1	Arend–Roland	1957 III	1956h	8 Apr. 1957	0.316	1.0002		2	5.5	
C/1957 P1	Mrkos	1957 V	1957d	1 Jul. 1957	0.355	0.9994		1	4.5	
C/1959 Y1	Burnham	1960 II	1959k	29 Mar. 1960	0.355	0.9994		4	8	
C/1961 R1	Humason	1962 VIII	1961e	10 Dec. 1962	2.13	0.990		6	1.5	
C/1965 S1	Ikeya–Seki	1965 VIII	1965f	21 Oct. 1965	0.007 79	0.999 92		–10	6	Sungrazer (Kreutz family)
C/1969 Y1	Bennett	1970 II	1969i	20 Mar. 1970	0.538	0.996		0.5		
C/1973 E1	Kohoutek	1973 XII	1973f	28 Dec. 1973	0.142	1.000		–3	6	International observation campaign
C/1975 V1	West	1976 VI	1975n	25 Feb. 1976	0.197	1.000		–3		
C/1980 E1	Bowell	1982 I	1980b	12 Mar. 1982	3.364	1.057				Active far from the Sun
C/1983 J1	Sugano–Saigusa–Fujikawa	1983 V	1983e	1 Apr. 1983	0.471	1.000				Approached Earth at 0.063 AU
C/1983 H1	IRAS–Araki–Alcock	1983 VII	1983d	21 May 1983	0.991	0.990				Approached Earth at 0.031 AU
C/1983 O1	Černis	1983 XII	1983l	21 Jul. 1983	3.33	1.002				Active at more than 20 AU
28P/	Neujmin 1	1984 XIX	1984c	8 Oct. 1984	1.553	0.776	18.2		8	
21P/	Giacobini–Zinner	1985 XIII	1984e	5 Sep. 1985	1.028	0.708	6.59		8.5	ICE space mission
1P/1982 U1	Halley	1986 III	1982i	9 Feb. 1986	0.587	0.967	75.99	2	2.5	Several space missions
C/1986 P1	Wilson	1987 VII	1986l	20 Apr. 1987	1.200	1.0003				
C/1988 Y1	Yanaka	1988 XXIV	1988r	11 Dec. 1988	0.43	1.000				Unusual composition
23P/1989 N1	Brorsen–Metcalf	1989 X	1989o	11 Sep. 1989	0.479	0.972	70.55		7.5	
29P/	Schwassmann–Wachmann 1	1989 XV		26 Oct. 1989	5.712	0.045	14.85			Nearly circular orbit
C/1989 XI	Austin	1990 V	$1989c_1$	10 Apr. 1990	0.350	1.0002				
C/1990 K1	Levy	1990 XX	1990c	24 Oct. 1990	0.939	1.0004				
43P/	Wolf–Harrington	1991 V		14 Apr. 1991	1.61	0.539	6.5		9	Unusual composition
103P/1991 N1	Hartley 2	1991 XV	1991t	11 Sep. 1991	0.953	0.719	6.26		7.5	
26P/	Grigg–Skjellerup	1992 XVIII		22 Jul. 1992	0.99	0.664	5.1		12	Giotto extended mission
109P/1992 S2	Swift–Tuttle	1992 XXVIII	1992t	12 Dec. 1992	0.958	0.963	135.0		4.5	
D/1993 F2	Shoemaker–Levy 9	1994 X	1993e							Broken by Jupiter and crashed into it
10P/	Tempel 2	1994 VII		16 Mar. 1994	1.483	0.552	5.47		9	
9P/	Tempel 1	1994 XIX	1993c	3 Jul. 1994	1.494	0.520	5.50		5.5	
6P/	d'Arrest			27 Jul. 1995	1.346	0.614	6.51		7	
73P/	Schwassmann–Wachmann 3			22 Sep. 1995	0.932	0.694	5.34		10	Nucleus split in 1995

continued

Table A.1. Famous comets (*continued*).

Comet (a)		(b)	(c)	Perihelion date	q (AU)	e	P (years)	m_{max}	H_0	Remark
122P/1995 S1	de Vico			6 Oct. 1995	0.659	0.962	74.36		7	
95P/	Chiron			14 Feb. 1996	8.454	0.385	50.73			Formerly listed as asteroid
C/1996 B2	Hyakutake			1 May 1996	0.230	0.9998	9000	0	5	Approached Earth at 0.10 AU
119P/1995 M2	Parker–Hartley			26 Jun. 1996	3.043	0.291	8.90		7	Formerly catalogued asteroid 1986 TF
22P/	Kopff			2 Jul. 1996	1.579	0.543	6.45		7	
46P/	Wirtanen			14 Mar. 1997	1.062	0.656	5.45		8	Target of Rosetta mission
C/1995 O1	Hale–Bopp			1 Apr. 1997	0.914	0.995	2400	−1	0	Extensive observational campaign
81P/	Wild 2			6 May 1997	1.58	0.539	6.39		7	Target of NEAR mission
2P/	Encke			23 May 1997	0.333	0.849	3.33		9	
49P/	Arend–Rigaux			12 Jul. 1998	1.368	0.611	6.61		11	Almost extinct

This table lists most of the comets quoted in the book plus several other comets of historical or scientific interest. The selection is subjective, and should not be used for statistical purposes. Comets are listed by order of perihelion date with their official designation (a). For comets that appeared before 1995, the old style designation (b) and provisional numbering (c) are also given. For short period comets, one of the last apparitions and/or historical returns is listed; thus they may be listed several times.

q: perihelion distance; e: eccentricity; P: orbital period; m_{max}: visual magnitude for maximum observed brightness (which depends upon visibility conditions and distance to Earth); H_0: visual magnitude reduced to 1 AU from Earth and Sun (this 'intrinsic magnitude' allows us to compare comets).

Table A.2. List of representative or peculiar asteroids.

Asteroid	a (AU)	e	P (years)	D (km)	Type	Family	Remark
Main-Belt asteroids							
(1) Ceres	2.768	0.076	4.60	913	G	Main Belt	
(2) Pallas	2.772	0.234	4.61	523	B	Main Belt	
(3) Juno	2.671	0.257	4.37	244	S	Main Belt	
(4) Vesta	2.362	0.090	3.63	501	V	Main Belt	
(243) Ida	2.863	0.044	4.84	33	S	Main Belt	Observed by Galileo
(951) Gaspra	2.210	0.173	3.29	15	S	Main Belt	Observed by Galileo
(2453) Mathilde	2.646	0.266	4.30	58		Main Belt	Observed by NEAR
(4979) Otawara	2.167	0.145	3.19	5.5		Main Belt	To be observed by Rosetta
(140) Siwa	2.732	0.217	4.51	110	P	Main Belt	To be observed by Rosetta
(153) Hilda	3.981	0.143	7.94	175	P	Hilda	
Earth-crossing asteroids							
(433) Eros	1.458	0.223	1.76	20	S	Apollo–Amor	To be observed by NEAR
(1862) Apollo	1.471	0.560	1.78	1.5	Q	Apollo–Amor	
(1566) Icarus	1.078	0.827	1.12	0.9		Apollo–Amor	
(4179) Toutatis	2.509	0.640	3.97	3		Apollo–Amor	Made close approaches to Earth
(2201) Oljato	2.174	0.712	3.20	1.8		Apollo–Amor	Possibly an extinct comet
(3200) Phaeton	1.271	0.890	1.43	6	F	Apollo–Amor	On a cometary orbit
(4015) Wilson–Harrington	2.644	0.622	4.30				Formerly classified as comet
Centaurs and distant asteroids							
(588) Achilles	5.201	0.147	11.79	147	DU	Trojan	
(944) Hidalgo	5.779	0.658	13.89	38	D		
(5335) Damocles	11.892	0.867	41.01				
(2060) Chiron	13.648	0.381	50.42	180	B	Centaur	Cometary activity
(5145) Pholus	20.226	0.571	90.96	(100)		Centaur	
(7066) Nessus	24.594	0.519	122.0			Centaur	
Trans-Neptunian objects							
1992 QB_1	44.298	0.077	294.8	(250)		Edgeworth–Kuiper belt	First discovered KBO
1993 FW	43.522	0.045	287.1	(250)		Edgeworth–Kuiper belt	
1993 RO	39.608	0.205	249.3	(150)		Edgeworth–Kuiper belt	'Plutino'
1996 TL_{66}	84.505	0.585	776.8	(550)			Atypical
Pluto	39.44	0.240	247.7	2300			For comparison

a: semi-major axis; e: eccentricity; P: orbital period; D: mean diameter.

Table A.3. Space missions to comets and asteroids.

Mission		Launch	Target	Date of encounter	
ICE	NASA	Aug. 1978	21P/Giacobini–Zinner	11 Sep. 1985	Flyby 7800 km
Vega 1	USSR	15 Dec. 1984	1P/Halley	6 Mar. 1986	Flyby 8900 km
Vega 2	USSR	21 Dec. 1984	1P/Halley	9 Mar. 1986	Flyby 8000 km
Giotto	ESA	2 Jul. 1985	1P/Halley	14 Mar. 1986	Flyby 600 km
			26P/Grigg–Skjellerup	10 Jul. 1992	Flyby $\approx$ 200 km
Sakigake	Japan	8 Jan. 1985	1P/Halley	11 Mar. 1986	Flyby 7×10^6 km
Suisei	Japan	18 Aug. 1985	1P/Halley	8 Mar. 1986	Flyby 0.15×10^6 km
Galileo	NASA/ESA	18 Oct. 1989	(951) Gaspra	29 Oct. 1991	Flyby
			(243) Ida + Dactyl	28 Aug. 1993	Flyby
NEAR	NASA	17 Feb. 1996	(253) Mathilde	27 Jun. 1997	Flyby
			(433) Eros	23 Nov. 1998	Flyby
				Feb. 2000	Rendezvous
DS1	NASA	24 Oct. 1998	(9969) Braille	29 Jul. 1999	Flyby
Stardust	NASA	7 Feb. 1999	81P/Wild 2	Mar. 2004	Dust sample return
SMART	ESA	2001	targets to be selected		Flyby
MUSES-C	NASA/Japan	Jan. 2002	(4660) Nereus	Sep. 2003	Sample return
Rosetta	ESA	Jan. 2003	(4979) Otawara	Jul. 2006	Flyby
			(140) Siwa	Jul. 2008	Flyby
			22P/Wirtanen	2012–2013	Rendezvous
Deep Impact	NASA	Jan. 2004	9P/Tempel 1	Jul. 2005	Impact on nucleus

Target objects prefixed by 'P' are comets. Those with numbers in brackets are asteroids.
This table was established in Summer 1999. All missions scheduled for after this date may be subject to modification or even cancellation.

Glossary

Accretion — A process in which matter accumulates to form massive objects, such as planetesimals, cometary nuclei, asteroids, planets or stars.

Active comet — A comet in which solar heating causes sublimation of ice and the formation of a coma.

Albedo — Fraction of incident light reflected by a body.

Aphelion — The point on an orbit which is furthest from the Sun.

Asteroid — An object of the Solar System, in orbit around the Sun, whose size lies between a fraction of one kilometre and 1000 km, and which exhibits no cometary activity.

Asteroid belt — The group of asteroids located between the orbits of Mars and Jupiter and which constitute the vast majority of known asteroids. These are also known as *Main Belt asteroids*.

Astronomical unit — One astronomical unit (AU) is the unit of length equal to the mean distance of the Earth from the Sun, namely about 150 million kilometres.

CCD — This stands for *charge coupled device*, meaning an array of photosensitive detectors read by a microprocessor. These are increasingly common in modern telescopes, where they replace the photographic plate.

Centaurs — Asteroids gravitating between Jupiter and Neptune, sometimes on highly eccentric orbits. One such, Chiron, is known to manifest cometary activity.

Circumstellar region — The immediate environment of a star in which its radiation and stellar wind interact with the rest of its protoplanetary disk and the local interstellar medium.

Coma — A roughly spherical envelope of gas and dust which develops around a cometary nucleus when it becomes active.

Cometary nucleus — The solid part of a comet, usually a few kilometres or a few dozen kilometres across, composed of ices and dust grains.

Cometary trail — Large cometary dust particles and debris which follow the comet's orbit.

Coronagraph An astronomical instrument containing a mask which creates an artificial eclipse, so that observations can be made close to a bright object such as the Sun, without dazzle.

Daughter molecule See parent molecule.

Eccentricity A parameter characterising the shape of an orbit, taking the value 0 for a circle, a value between 0 and 1 for an ellipse, the value 1 for a parabola, and a value greater than 1 for a hyperbola.

Ecliptic The great circle of the Sun's apparent path on the celestial sphere.

Edgeworth–Kuiper belt Often referred to as the Kuiper belt. A group of planetoids (asteroids or cometary nuclei) gravitating close to the ecliptic beyond Neptune. It is suggested that these supply the group of short period comets in the Jupiter family.

Galactic nebula An interstellar cloud which shines either by emission from its gas, or by reflection of starlight from its dust.

Halley family A family of comets whose periods lie between 20 and 200 years, and whose orbital inclinations are uniformly distributed relative to the ecliptic.

Heliocentric distance Distance from the Sun, usually given in astronomical units.

Infrared The region of the electromagnetic spectrum defined by wavelengths lying between 100 μm and 800 nm. Apart from a few 'windows', accessible at well-placed, high altitude sites, observations in this region must be made from planes, balloons and satellites.

Interstellar clouds Clouds of gas and dust which, together with the stars, constitute a major part of a galaxy. Low density clouds, composed mainly of atoms, are called *diffuse clouds*, whereas high density clouds are referred to as *dense clouds*, and clouds shrouded by dust, which blocks out stellar radiation, are called *dark clouds*. Dense, dark clouds are rich in molecules, and these are called *molecular clouds*.

Interstellar grains Small solid particles found in interstellar clouds and interstellar space, which may have been incorporated directly into cometary nuclei.

Ionosphere In cometary terminology, this refers to the inner region of the coma, bounded by the *ionopause*, in which cometary ions are not perturbed by the solar wind and the magnetic field is zero.

Isotope Atom of a given element having the same number of protons and electrons, and hence the same chemical properties, but a different number of neutrons. *Relative isotopic abundances* are an important indicator in theories about the formation of the Universe.

Jets The inner coma of a comet is rarely symmetrical. Jets consist of gases sublimed from nuclear ices, and dust carried away with them. They are produced in *active regions*, heated up by the Sun.

Jupiter family A family of comets of period about 6 years whose orbits lie close to the plane of the ecliptic. They are strongly influenced by Jupiter's gravitational attraction, which perturbs their orbits.

Kelvin Unit of absolute temperature, denoted by K. *Absolute zero* 0 K is the lowest temperature that could ever be attained, according to the principles of thermodynamics; it corresponds to –273.15 °C, and 0 °C corresponds to 273.15 K.

Kreutz family A family of comets, also known as the *sungrazing comets* or *sungrazers*, whose orbits are closely grouped and bring them very near the Sun (at little more than one solar radius).

Magnetosheath or cometosheath The region enclosed between the shock wave (*bow shock*) and the *ionopause* (or *cometopause*), which is the scene of intense interaction between cometary ions and ions in the solar wind.

Magnitude A measure of the luminosity of an object, on a logarithmic scale. It increases by five units as the luminosity decreases by a factor of 100. It can be defined for different colours. The *visual magnitude* refers to the luminosity at wavelengths within the range of sensitivity of the eye.

Meteor or shooting star The luminous phenomenon seen when a meteoroid enters the Earth's atmosphere. A group of meteors which appear to radiate from a common point in the sky is called a *meteor shower.*

Meteorite An object of extraterrestrial origin which has survived passage through the atmosphere and reached the surface of the Earth.

Meteoroid A small body in the Solar System, such as a particle of cometary dust. Meteoroids in the form of small rocks may become meteors or meteorites if they collide with the Earth's atmosphere.

Nanotesla A unit of magnetic field, sometimes referred to as 'gamma' by geophysicists, in terms of which the terrestrial magnetic field has value around 50 000 nT in northern latitudes.

Neutral sheet In a cometary ion tail, this is the surface separating two lobes in which the magnetic field has opposite polarity.

New comet The term designating a comet which has just escaped from the Oort cloud and whose first passage through the inner Solar System is being observed.

Non-gravitational forces In cometology, any force on the cometary nucleus which is not due to the gravitational attraction of the Sun or the planets, such as the reaction force associated with emission of gases in some favoured direction.

Nucleosynthesis The atomic processes by which heavy atoms are formed from lighter atoms in stellar cores.

Oort cloud A hypothetical cloud of comets, uniformly distributed over distances between 20 000 and 100 000 km from the Sun, proposed as the source of new comets.

Organic compound A molecule made up of the elements carbon and hydrogen and possibly also oxygen and nitrogen.

Osculating parabola The parabola best fitting the orbit of a new comet whose eccentricity has not yet been determined.

Parent molecule A molecule produced directly by sublimation of cometary ice, to be contrasted with *secondary products* produced by dissociation of parent molecules under solar UV irradiation. These secondary products, often inaccurately referred to as *daughter molecules*, may be radicals (molecular fragments unstable in laboratory conditions), atoms, atomic ions or molecular ions.

Perihelion The point of an orbit which is closest to the Sun.

Period The *orbital period* is the time taken by an orbiting object to complete one orbit. The *rotation period* is the time taken by a body to complete one turn about its axis.

Periodic comet These are divided into those of *short period*, less than 200 years, and those of *long period*, greater than 200 years, which are difficult to observe on more than one occasion.

Photometry Any technique for measuring the intensity of radiation.

Planetesimals Any small bodies in the primordial solar nebula, which could accrete to form cometary nuclei, asteroids and planets.

Primordial solar nebula A cloud of gas and dust which surrounded the protosun during its formation. Having assumed the shape of a flattened disk, this cloud gave birth to planetesimals.

Radio waves The radio region of the electromagnetic spectrum includes all wavelengths greater than 100 μm. Subdivisions are the *millimetre range* from 1 to 10 mm, and the *submillimetre range* from 0.1 to 1 mm. Part of the latter range is inaccessible to Earth-based radiotelescopes.

Radical See parent molecule.

Refractory molecule A molecule, such as those in cometary dusts, which can remain in the solid state even at relatively high temperatures.

Resolving power (resolution) A spectrometer of resolving power $\lambda/\delta\lambda$ can distinguish wavelengths λ and $\lambda + \delta\lambda$. A mass spectrometer of resolution $m/\delta m$ can distinguish masses m and $m + \delta m$.

Solar and stellar winds A continuous flow of matter escaping from the Sun or a star. The solar wind is essentially made up of ions, travelling at speeds up to 400 km/s.

Swings effect The variation in the structure of cometary spectra caused by Doppler shifting of the solar spectrum which excites cometary molecules.

Tail Comets may possess several types of tail. The *dust tail* is caused by solar radiation pressure on cometary dusts, and the *ion tail* is due to the interaction of ions in the solar wind with ions from the comet.

Trans-Neptunian objects This is another name for the objects in the Edgeworth–Kuiper belt.

Trojan asteroids A family of asteroids which share Jupiter's orbit, remaining roughly equidistant from both Jupiter and the Sun.

Ultraviolet (UV) radiation The region of the electromagnetic spectrum defined by wavelengths lying between about 10 and 350 nm. Down to 300 nm, the UV can still be observed from the Earth, although it is attenuated by the atmosphere. At shorter wavelengths, sounding rockets and observation satellites are required.

Visible wavelengths The visible region of the electromagnetic spectrum consists of those wavelengths between about 350 and 800 nm, which can be detected by the human eye.

Volatile molecules Molecules, such as those in cometary ices, which sublime or condense at relatively low temperatures.

Bibliography

General articles and monographs

[1] C. Arpigny, J. Rahe *et al.*, *Atlas of Cometary Spectra*, Kluwer, Dordrecht (1999) (in press).

[2] M.C. Festou, H. Rickman and R.M. West, Comets, *Astronomy and Astrophysics Reviews*, **4**, 363–447, and **5**, 37–163 (1993).

[3] D.W. Green (editor), *International Comet Quarterly*, Smithsonian Astrophysical Observatory, Cambridge, USA.

[4] M. Grewing, F. Praderie and R. Reinhard (editors), Exploration of Halley's Comet, *Astronomy and Astrophysics*, **187** (1987).

[5] W.F. Huebner (editor), *Physics and Chemistry of Comets*, Springer-Verlag, Berlin (1990).

[6] K.S. Krishna Swamy, *Physics of Comets* (2nd ed.), World Scientific, Singapore (1997).

[7] B.G. Marsden and G.V. Williams, *Catalog of Cometary Orbits* (13th ed.), IAU, Central Bureau for Astronomical Telegrams, Cambridge, Massachusetts (1999).

[8] J. Mason (editor), *Comet Halley, Investigations, Results, Interpretations*, Ellis Horwood, New York (1990).

[9] M.J. Mumma, P.R. Weissman, S.A. Stern, Comets and the Origin of the Solar System: Reading the Rosetta Stone. In: *Protostars and Planets III*, E.H. Levy and J.I. Lunine (editors), University of Arizona Press, Tucson, 1177–1252 (1993).

[10] R.L. Newburn, M. Neugebauer and J. Rahe (editors), *Comets in the Post-Halley Era*, Kluwer, Dordrecht (1991).

[11] K.S. Noll, H.A. Weaver, P.D. Feldman, *The Collision of Comet Shoemaker–Levy 9 and Jupiter*, Cambridge University Press, Cambridge (1996).

[12] J.R. Spencer, J. Mitton, *The Great Comet Crash*, Cambridge University Press, Cambridge (1995).

[13] L.L. Wilkening (editor), *Comets*, University of Arizona Press, Tucson (1982).

[14] D.K. Yeomans, *Comets: A Chronological History of Observation, Science, Myth and Folklore*, John Wiley and sons, New York (1991).

Articles of historical interest

[15] F. Arago, *Les comètes*, re-edition of an extract from *L'Astronomie populaire*, 1858, Blanchard, Paris (1986).

[16] L. Biermann, Kometenschweife und Solar Korpuskularstrahlung, *Z. Astrophys.*, **29**, 274–286 (1951).

[17] A.H. Delsemme and P. Swings, Hydrates de gaz dans les noyaux cométaires et les grains interstellaires, *Ann. Astrophys.*, **15**, 1–6 (1952).

[18] G.P. Kuiper, On the Origin of the Solar System, included in J.A. Hynek (editor), *Astrophysics*, McGraw–Hill, New York, 357–424 (1951).

[19] *Nature* (special issue), Encounters with comet Halley. The first results, **321** (1986).

[20] J.H. Oort, The structure of the cloud of comets surrounding the Solar System and a hypothesis concerning its structure, *Bull. Astr. Inst. Netherl.*, **11**, 91–110 (1950).

[21] F.L. Whipple, A comet model I: the acceleration of comet Encke, *Astrophys. J.*, **111**, 375–395 (1950).

[22] F.L. Whipple, A comet model II: physical relations for comets and meteors, *Astrophys. J.*, **113**, 464–474 (1951).

Other references

[23] C. Arpigny, Spectroscopie cométaire, *J. de Physique*, supplement to issue 10, **32**, C5a129–C5a141 (1971).

[24] J.N. Bahcall, L. Spitzer, The Space Telescope, *Scientific American*, July 1982, 40.

[25] H. Balsiger, H. Fechtig, J. Geiss, A close look at Halley's comet, *Scientific American*, September 1988, 96.

[26] F. Bass, T.R. Geballe, D.M. Walther, Spectroscopy of the 3.4 micron emission feature in comet Halley, *Astrophys. J.*, **311**, L97–L102 (1986).

[27] F. Biraud, G. Bourgois *et al.*, OH observations of comet Kohoutek (1973f) at 18 cm wavelength, *Astron. Astrophys.*, **34**, 163–166 (1974).

[28] N. Biver, D. Bockelée-Morvan, P. Colom *et al.*, Long-term evolution of the outgassing of comet Hale–Bopp from radio observations. *Earth, Moon and Planets*, in press (1999).

[29] D. Bockelée-Morvan, J. Crovisier, The role of water in the thermal balance of the coma, ESA SP-278, 253–240 (1987).

[30] D. Bockelée-Morvan, J. Crovisier, Les molécules des comètes, *La Recherche*, **271**, 1272–1278 (1994).

[31] D. Bockelée-Morvan, T.Y. Brookes, J. Crovisier, On the origin of the 3.2–3.6 μm emission feature in comets, *Icarus*, **116**, 18–39 (1995).

[32] D. Bockelée-Morvan, D. Gautier, D.C. Lis *et al.*, Deuterated water in comet C/1996 B2 (Hyakutake) and its implications for the origins of comets, *Icarus*, **133**, 147–162 (1998).

[33] J.C. Brandt, M.B. Nieder, The structure of cometary tails, *Scientific American*, January 1986, 48.

[34] T.Y. Brooke, A.T. Tokunaga, H.A. Weaver *et al.*, Detection of acetylene in the infrared spectrum of comet Hyakutake, *Nature*, **383**, 606–608 (1996).

[35] M. Combes, V.I. Moroz *et al.*, The 2.5–12 μm spectrum of comet Halley from the IKS-Vega experiment, *Icarus*, **76**, 404–436 (1988).

[36] G. Cremonese, H. Boehnhardt, J. Crovisier *et al.*, Neutral sodium from comet Hale–Bopp: a third type of tail, *Astrophys. J.*, **490**, L199–L202 (1997).

[37] J. Crovisier, Molecular abundances in comets. In: *Asteroids, Comets, Meteors 1993*, A. Milani *et al.* (editors), Kluwer, Dordrecht, pp. 313–326 (1993).

[38] J. Crovisier, N. Biver *et al.*, Carbon monoxide outgassing from comet P/Schwassmann–Wachmann 1, *Icarus*, **115**, 213–216 (1995).

[39] J. Crovisier, K. Leech, D. Bockelée-Morvan *et al.*, The spectrum of comet Hale–Bopp (C/1995 O1) observed with the Infrared Space Observatory, *Science*, **275**, 1904–1907 (1997).

[40] J.K. Davies, S.F. Green, T.R. Geballe, The detection of a strong 3.28 μm emission feature in comet Levy, *Mon. Not. R. Astron. Soc.*, **251**, 148–151 (1991).

[41] A.H. Delsemme, Chemical composition of cometary nuclei. In: *Comets*, L.L. Wilkening (editor), University of Arizona Press, Tucson, pp. 85–130 (1982).

[42] D. Despois, J. Crovisier, Observations of hydrogen cyanide in comet Halley, *Astron. Astrophys.*, **160**, L11–L12 (1986).

[43] D. Despois, D. Bockelée-Morvan *et al.*, Radio line observations of comet P/Swift–Tuttle 1992t at IRAM, *Planet. Space Sci.,* **44**, 529–539 (1996).

[44] B. Donn, The accumulation and structure of comets. In: *Comets in the Post-Halley era*, R.L. Newburn *et al.* (editors), Kluwer, Dordrecht, pp. 335–359 (1991).

[45] U. Fink, Comet Yanaka (1998r): a new class of carbon-poor comet, *Science*, **257**, 1926–1929 (1992).

[46] J.M. Greenberg, The structure and evolution of interstellar grains, *Scientific American*, June 1984, 124.

[47] J.M. Greenberg, The evidence that comets are made of interstellar dust. In: *Comet Halley, Investigations, Results, Interpretations*, J. Mason (editor), Ellis Horwood, New York, pp. 99–120 (1990).

[48] K.I. Gringauz, T.I. Gombosi *et al.*, First *in situ* plasma and neutral gas measurements at comet Halley, *Nature*, **321**, 282–285 (1986).

[49] O.R. Hainaut, Selected observational studies of minor bodies in the Solar System, *Thesis*, Liège University (1994).

[50] J.K. Harmon, S.J. Ostro, L.A.M. Benner *et al.*, Radar detection of the nucleus and coma of comet Hyakutake (C/1996 B2), *Science*, **278**, 1921–1924 (1997).

[51] C. Hayashi, K. Nakazawa, Y. Nakagawa, Formation of the Solar System. In: *Protostars and Planets II*, D.C. Black, M.S. Matthews (editors), University of Arizona Press, Tucson, pp. 1100–1153 (1985).

[52] D.W. Hughes, Cometary magnitude distribution and the fading of comets, *Nature*, **325**, 221–232 (1987).

[53] D. Jewitt, J. Luu, C. Trujillo, Large Kuiper Belt objects: the Mauna Kea 8K CCD survey, *Astron. J.*, **115**, 2125–2135 (1998).

[54] D.C. Jewitt, J.X. Luu, A CCD portrait of comet P/Tempel 2, *Astron. J.*, **97**, 1766–1790 (1989).

[55] D.C. Jewitt, J.X. Luu, The Solar System beyond Neptune, *Astron. J.*, **109**, 1867–1876 (1995).

[56] L. Kamèl, The comet light curve atlas, *Astron. Astrophys. Suppl.*, **92**, 85–149 (1992).

[57] J. Kissel, R.Z. Sagdeev *et al.*, Composition of comet Halley dust particles from Vega observations, *Nature*, **321**, 280–282 (1986).

[58] V.A. Krasnopolsky, M.J. Mumma *et al.*, Detection of soft X-rays and a sensitive search for noble gases in comet Hale–Bopp (C/1995 O1), *Science* **277**, 1488–1491 (1997).

[59] L. Kresàk, Comets, existing populations. In: *Asteroids, Comets, Meteors 1993*, A. Milani *et al.* (editors), Kluwer, Dordrecht, pp. 77–94 (1994).

[60] H.P. Larson, H.A. Weaver *et al.*, Airborne infrared spectroscopy of comet Wilson (1986l) and comparison with comet Halley, *Astrophys. J.*, **338**, 1106–1114 (1989).

[61] R.S. Lewis, E. Anders, Interstellar matter in meteorites, *Scientific American*, August 1983, 66.

[62] D.C. Lis, D. Mehringher, D. Benford *et al.*, New molecular species in comet C/1995 O1 (Hale–Bopp) observed with the Caltech Submillimeter Observatory. *Earth, Moon and Planets* (in press) (1999).

[63] C.M. Lisse, K. Dennerl *et al.*, Discovery of X-ray and extreme ultraviolet emission from comet C/Hyakutake 1996 B2, *Science* **274**, 205–209 (1996).

[64] D.J. Malaise, Collisional effects in cometary atmospheres, *Astron. Astrophys.*, **5**, 209–227 (1970).

[65] B.G. Marsden, The sungrazing comet group, *Astron. J.*, **72**, 1170–1182 (1967).

[66] G.J. Moreels, J. Clairemidi *et al.*, Detection of a polycyclic aromatic molecule in comet P/Halley, *Astron. Astrophys.*, **282**, 643–656 (1994).

[67] D. Morrison, *The Spaceguard Survey*, NASA-JPL Report (1992).

[68] M.J. Mumma, M.A. DiSanti, N. Dello Russo *et al.*, Detection of abundant ethane and methane, along with carbon monoxide and water, in comet

C/1996 B2 (Hyakutake): evidence for interstellar origin, *Science*, **272**, 1310–1314 (1996).

[69] M.J. Mumma, H.A. Weaver *et al.*, Detection of water vapor in Halley's comet, *Science*, **232**, 1523–1528 (1986).

[70] E.P. Ney, Optical and infrared observations of bright comets in the range 0.5 μm to 20 μm. In: *Comets*, L.L. Wilkening (editor), University of Arizona Press, Tucson, pp. 323–340 (1982).

[71] W.H. Osborn, M.F. A'Hearn *et al.*, Standard stars for photometry of comets, *Icarus*, **88**, 228–245 (1990).

[72] A. Owens, A.N. Parmar *et al.*, Evidence for dust-related X-ray emission from comet C/1995 O1 (Hale–Bopp), *Astrophys. J.*, **493**, L47–L51 (1998).

[73] E. Pantin, P.O. Lagage, Des traces de planètes autour de β Pictoris, *Pour la Science*, **209**, 20 (1995).

[74] C. Sagan, *Cosmos*, Mazarine, Paris.

[75] D.G. Schleicher, Comet taxonomy and evolution. In: *Asteroids, Comets, Meteors 1993*, A. Milani *et al.* (editors), Kluwer, Dordrecht, pp. 415–428 (1993).

[76] D.G. Schleicher, S.J. Bus, D.J. Osip, The anomalous molecular abundances of comet P/Wolf–Harrington, *Icarus*, **104**, 157–166 (1993).

[77] Z. Sekanina, S.M. Larson *et al.*, Major outburst of comet Halley at a heliocentric distance of 14 AU, *Astron. Astrophys.*, **263**, 367–382 (1992).

[78] M.C. Senay, D. Jewitt, Coma formation driven by carbon monoxide release from comet Schwassmann–Wachmann I, *Nature*, **371**, 229–231 (1994).

[79] H.A. Weaver, P.D. Feldman, Probing the nature of comets with the HST. In: *ESO Conference and Workshop Proceedings*, **44**, P. Benvenutti, E. Schreier (editors), Sardinia, Italy (1992).

[80] H.A. Weaver, M.F. A'Hearn *et al.*, The Hubble Space Telescope observing campaign on comet P/Shoemaker-Levy 9 (1993e), *Science*, **267**, 1282–1288 (1995).

[81] D.C.B. Whittet, W.A. Schatte, A.G.G.M. Tielens *et al.*, An ISO SWS view of interstellar ices: first results, *Astron. Astrophys.*, **315**, L357–L360 (1996).

[82] T. Yamamoto, Formation environment of cometary nuclei in the primordial solar nebula, *Astron. Astrophys.*, **142**, 31–36 (1985).

Internet Addresses

[83] http://cfa-www.harvard.edu/iau/cbat.html The Central Bureau for Astronomical Telegrams. Recent news on comets and other transient astronomical events; access to IAU circulars.

[84] http://cfa-www.harvard.edu/iau/icq.html The International Comet Quarterly. Recent measurements of cometary total visual magnitudes, mainly by amateurs.

[85] http://encke.jpl.nasa.gov/index.html and http://encke.jpl.nasa.gov/RecentObs.html (maintained by Charles S. Morris) also give comprehensive reports of cometary observations (magnitude measurements and images) by amateurs.

All these sites have links with many other internet sites relevant to comets.

Index